Springer Theses

Recognizing Outstanding Ph.D. Research

Aims and Scope

The series "Springer Theses" brings together a selection of the very best Ph.D. theses from around the world and across the physical sciences. Nominated and endorsed by two recognized specialists, each published volume has been selected for its scientific excellence and the high impact of its contents for the pertinent field of research. For greater accessibility to non-specialists, the published versions include an extended introduction, as well as a foreword by the student's supervisor explaining the special relevance of the work for the field. As a whole, the series will provide a valuable resource both for newcomers to the research fields described, and for other scientists seeking detailed background information on special questions. Finally, it provides an accredited documentation of the valuable contributions made by today's younger generation of scientists.

Theses are accepted into the series by invited nomination only and must fulfill all of the following criteria

- They must be written in good English.
- The topic should fall within the confines of Chemistry, Physics, Earth Sciences, Engineering and related interdisciplinary fields such as Materials, Nanoscience, Chemical Engineering, Complex Systems and Biophysics.
- The work reported in the thesis must represent a significant scientific advance.
- If the thesis includes previously published material, permission to reproduce this must be gained from the respective copyright holder.
- They must have been examined and passed during the 12 months prior to nomination.
- Each thesis should include a foreword by the supervisor outlining the significance of its content.
- The theses should have a clearly defined structure including an introduction accessible to scientists not expert in that particular field.

More information about this series at http://www.springer.com/series/8790

Keisuke Sugiura

Development of a Numerical Simulation Method for Rocky Body Impacts and Theoretical Analysis of Asteroidal Shapes

Doctoral Thesis accepted by
Nagoya University, Nagoya, Japan

Springer

Author
Dr. Keisuke Sugiura
Earth-Life Science Institute
Tokyo Institute of Technology
Meguro, Tokyo, Japan

Supervisor
Prof. Shu-ichiro Inutsuka
Department of Physics
Graduate School of Science
Nagoya University
Nagoya, Japan

ISSN 2190-5053 ISSN 2190-5061 (electronic)
Springer Theses
ISBN 978-981-15-3721-9 ISBN 978-981-15-3722-6 (eBook)
https://doi.org/10.1007/978-981-15-3722-6

This Springer imprint is published by the registered company Springer Nature Singapore Pte Ltd.
The registered company address is: 152 Beach Road, #21-01/04 Gateway East, Singapore 189721, Singapore

Supervisor's Foreword

Asteroids are rocky objects smaller than planets mainly found in the inner Solar System and are rich in variety of their geometrical shapes. Although some are nearly ellipsoidal, many objects are known to be far from sphere. Some objects even look like sea-otters. Such enigmatic shapes of asteroids stimulate our interest on their origin. The explanation for the shape of each object may include the formation process in the initial phase, the secular evolution after the formation, and accidental events happened later in its evolution. Among them, the understanding of the formation of the initial shape should be the starting point for further studies. The formation processes of the asteroids are supposed to be the collisions of planetesimals that are rocky objects much smaller than planets and supposed to be created in the early phase of the solar nebula. Since the experimental study of the collisions of such astronomical objects is impractical, numerical experiments should play crucial roles in our investigation for the formation processes. However, actual numerical calculation for the collision of rocky objects requires accurate solution of elastic dynamics equation in fully three-dimensional geometry and description of friction, fracture, and destruction, which is not yet well developed and challenging for frontier of theoretical research.

In this thesis, Keisuke Sugiura has developed numerical algorithm based on Lagrangian particles for solving elastic dynamics equations in three-dimensional geometry and implemented methods to describe friction, fracture, and destruction. In effect this work opens a new theoretical research field for the formation and evolution of asteroids. Indeed, he is now solving many interesting problems, such as the statistical distributions of various shapes of asteroids, the possible formation mechanism of highly elongated asteroid "1I/'Oumuamua" that is supposed to have come from outside of the Solar System, and the reason for the unique top shape of "Ryugu".

As applications of his method, this thesis includes only the subjects related to asteroids. However, potential application of his method on the collision of planetesimals with various sizes in more general context may provide a significant impact on the theory of planet formation, because planetesimals are building blocks of rocky planets and asteroid-size objects are regarded as the precursors of the rocky planets. After the discovery of the first exo-planets in 1995, the scientific community keeps increasing and the various researches on planet formation processes are developing rapidly but our understanding of the basic consequence of the collision of planetesimals and rocky planets are limited. I hope this thesis provides a useful step to accelerate our understanding of the planet formation processes.

Nagoya, Japan
January 2020

Prof. Shu-ichiro Inutsuka

Parts of this thesis have been published in the following journal articles:

1. K. Sugiura and S. Inutsuka, “An extension of Godunov SPH: Application to negative pressure media,” J. Comput. Phys. **308** (2016) 171–197, [https://doi.org/10.1016/j.jcp.2015.12.030]
2. K. Sugiura and S. Inutsuka, “An extension of Godunov SPH II: Application to elastic dynamics,” J. Comput. Phys. **333** (2017) 78–103, [https://doi.org/10.1016/j.jcp.2016.12.026]
3. K. Sugiura, H. Kobayashi and S. Inutsuka, “Toward understanding the origin of asteroid geometries. Variety in shapes produced by equal-mass impacts,” Astron. Astrophys. **620** (2018) A167, [https://doi.org/10.1051/0004-6361/201833227]
4. K. Sugiura, H. Kobayashi and S. Inutsuka, “Collisional Elongation: Possible Origin of Extremely Elongated Shape of 1I/‘Oumuamua,” Icarus **328** (2019) 14–22, [https://doi.org/10.1016/j.icarus.2019.03.014]
5. K. Sugiura, H. Kobayashi and S. Inutsuka, “High-Resolution Simulations of Catastrophic Disruptions: Resultant Shape Distributions,” Planet. Space Sci. **181** (2020) 104807, [https://doi.org/10.1016/j.pss.2019.104807]

Acknowledgements

First of all, I would like to show my greatest appreciation to Shu-ichiro Inutsuka for his guidance regarding the direction of my researches. I would also like to express my great gratitude to Hiroshi Kobayashi for giving me very detailed comments on our papers and knowledge about planet formation and celestial mechanics.

I am deeply grateful to all researchers who helped and encouraged my works. I would like to thank Hidekazu Tanaka for giving me useful comments when I visited Tohoku University, Takeru Suzuki, Kazunari Iwasaki, and Yusuke Tsukamoto for the discussion and helps regarding the numerical simulation method and coding technique, Ryuki Hyodo for useful comments on our third paper, Natsuki Hosono for helping me to parallelize our simulation code using FDPS, Ryo Suetsugu for giving me the motivation to start the research about 1I/'Oumuamua, and Martin Jutzi for giving me the opportunity to introduce my researches at the University of Bern.

I also thank all (ex-)members of the Ta laboratory in Nagoya University, especially Yuri Fujii and Kenji Kurosaki for staying in the same room and encouraging my daily works.

I have been supported by a Grant-in-Aid of Japan Society for the Promotion of Science under KAKENHI Grant Number JP 17J01703. Numerical simulations in this work were carried out on Cray XC30 and Cray XC50 at the Center for Computational Astrophysics, National Astronomical Observatory of Japan.

Finally, I would like to express my gratitude to my father and mother for supporting me throughout my life, and I also thank to my gymnastics coaches, Shojiro Ishida, Shunsuke Masaaki, Yumi Kato, and Shun-ichi Matsumoto for nurturing my personality through gymnastics.

Contents

Chapter 1
Introduction

Abstract Our planetary system is formed through collisions between bodies with various sizes, that is, micrometer-sized dust particles, km-sized planetesimals, and 1,000 km-sized protoplanets. Bodies grow through collisional coalescence of bodies, and they eventually become the planets in the present solar system. Thus, the collision is one of the most important processes for the formation of the solar system. The present solar system has not only the planets but also smaller asteroids. Since asteroids do not experience significant melting thanks to those small sizes, we can obtain information about the past environment of the solar system from asteroids. Especially, shapes of asteroids tell us what kind of collisions they experienced in the past solar system because shapes of asteroids are mainly altered by collisions between asteroids and resultant shapes depend on impact conditions. In this chapter, we introduce the overview of the solar system, the properties of asteroids, the standard solar system formation scenario, and the collisional phenomena between rocky minor bodies as basic knowledge to extract information of the past solar system from asteroids' shapes.

Keywords Solar system · Asteroids · Solar system formation scenario · Collisions between rocky bodies

1.1 Asteroids in the Solar System

Our solar system is mainly composed of the planets orbiting around the sun. The solar system has eight planets, that is, Mercury, Venus, the Earth, Mars, Jupiter, Saturn, Uranus, and Neptune. They are the major bodies orbiting around the sun and account for the almost all mass of the solar system. However, not only the planets but also other minor bodies move around the sun. The minor objects include asteroids, comets, and trans-neptunian objects.

The number of the minor bodies in the solar system is clearly much larger than that of the major bodies. Especially, great number of asteroids are found so far.

K. Sugiura, *Development of a Numerical Simulation Method for Rocky Body Impacts and Theoretical Analysis of Asteroidal Shapes*, Springer Theses,
https://doi.org/10.1007/978-981-15-3722-6_1

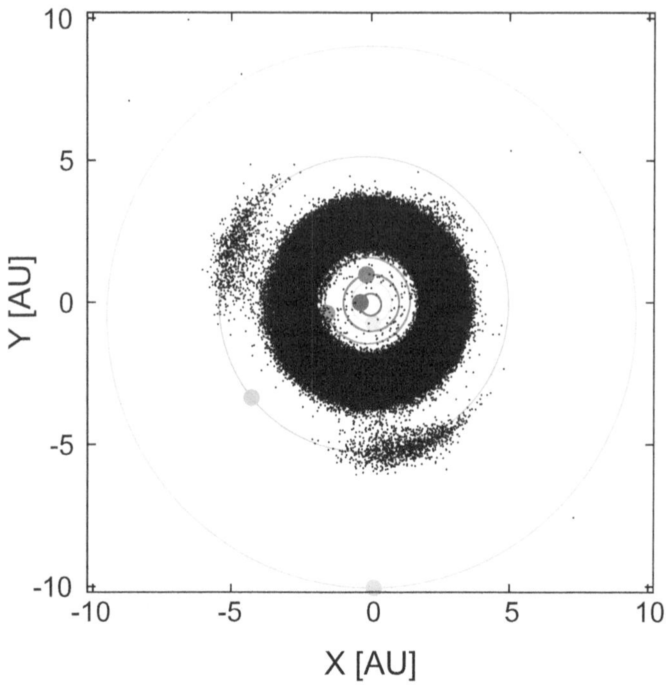

Fig. 1.1 Positions of the inner six planets and asteroids with diameters larger than 1 km at 00:00:00 on 1st January, 2018. The orbits of the planets are indicated by the solid curves. The orbit of the Earth is on the XY plane, and X axis corresponds to the direction of the vernal equinox. The positions and the orbits are calculated from orbital elements of the planets and the asteroids obtained from Small Body Database Search Engine of Jet Propulsion Laboratory. The blue point and solid curve show the position and orbit of Mercury, the yellow ones show those of Venus, the brown ones show those of the Earth, the red ones show those of Mars, the green ones show those of Jupiter, and the gray ones show those of Saturn. The black dots show the positions of the asteroids

According to Small Body Database Search Engine of Jet Propulsion Laboratory,[1] the number of all asteroids with known orbits is about 800,000. Even the number of the asteroids with diameters larger than 1 km is more than 100,000. The largest asteroid (1) Ceres, which was discovered by Giuseppe Piazzi on 1801, has the diameter of about 1,000 km (Note that (1) Ceres is no longer an asteroid but now classified to a dwarf planet). Thus more than 100,000 asteroids with diameters of 1–1,000 km are observed, and we can achieve statistical information regarding to properties of the solar system from asteroids.

Figure 1.1 shows the positions of the inner six planets and asteroids with diameters larger than 1 km. It is clear at a glance that asteroids mainly exist between the orbits of Mars and Jupiter. The region where asteroids mainly exist is called the main belt, and the asteroids in this region are called main-belt asteroids. We also notice that there

[1] https://ssd.jpl.nasa.gov/sbdb_query.cgi#x.

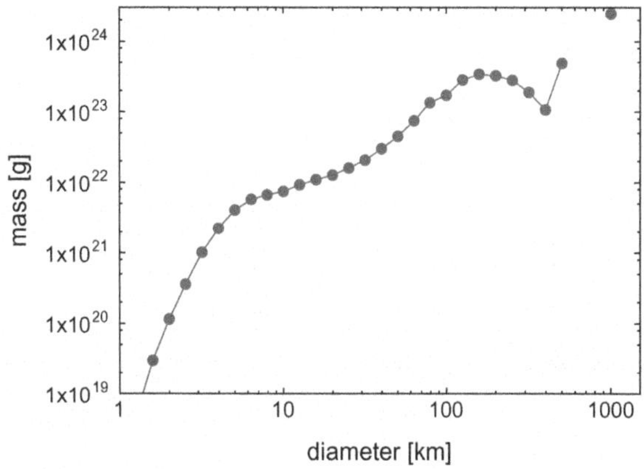

Fig. 1.2 Incremental mass distribution of asteroids calculated from diameters of asteroids obtained from Small Body Database Search Engine of Jet Propulsion Laboratory. Here, we assume that the density of asteroids is $3\,\mathrm{g/cm^3}$. Each point shows the total mass of asteroids with diameters between $10^{0.05}D_c - 10^{-0.05}D_c$, where D_c shows the center of each diameter bin. The horizontal axis represents diameters of asteroids, and the vertical axis represents the total mass of asteroids with diameters in each diameter bin

are two groups of asteroids existing in the Lagrange points of Jupiter. These asteroids are called Jupiter trojans. There are some asteroids that have orbits close to that of the Earth. These asteroids are called near-Earth asteroids, which include the asteroids Itokawa and Ryugu visited by the spacecraft Hayabusa and Hayabusa2. Almost all near-Earth asteroids have diameters less than 10 km, and they are considered to be originated from the main belt through the migration due to the Yarkovsky effect or the scattering due to the planets (e.g., Binzel et al. 1992; Michel et al. 2005).

Figure 1.2 shows the incremental mass distribution of asteroids. Masses of asteroids are estimated from those diameters obtained from Small Body Database Search Engine of Jet Propulsion Laboratory. From Fig. 1.2, we notice that the total mass of asteroids with diameters in each diameter bin increases with increasing diameters of asteroids. We also notice that there is a peak of the mass around the diameter $\approx$100 km (see also Bottke et al. 2005; Kobayashi et al. 2016). Note that the two diameter bins with the diameter >500 km in Fig. 1.2 include only three asteroids (Ceres, Pallas, and Vesta), and these points are statistically insignificant. Thus, asteroids with the diameter $\approx$100 km account for large fraction of the total mass of asteroids.

Shapes of the planets or large satellites such as the moon are almost complete spheres. However, as we notice from the sea-otter shape of the asteroid Itokawa, many asteroids have irregular shapes that are different from spherical shapes. This is because material strength is dominant rather than the self gravity for small bodies (Hughes and Cole 1995; Lineweaver and Norman 2010). Shapes of asteroids are mainly measured using light curves, that is, luminosity variation of asteroids in time. Rotations of asteroids vary cross sections of asteroids with respect to observers, and this causes luminosity changes. Kaasalainen and Torppa (2001) and Kaasalainen

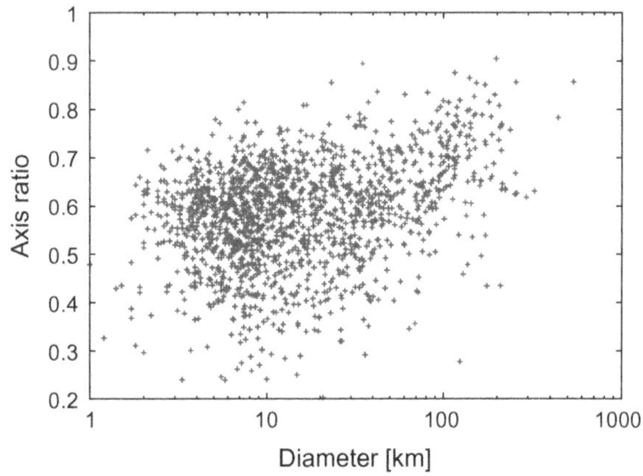

Fig. 1.3 Axis ratios of asteroids obtained from shape models stored in Database of Asteroid Models from Inversion Technique. The horizontal axis represents diameters of asteroids, and the vertical axis represents the ratios of the minor to major axis lengths of asteroids. The axis ratios are measured through bottom-up method (see Sect. 3.2)

et al. (2001) provide the novel technique that predicts shapes of asteroids from those light curves. Shape models of asteroids obtained from light curves are summarized in Database of Asteroid Models from Inversion Technique[2] (Ďurech et al. 2010). Shapes of asteroids are also measured by in-situ observations (e.g., Fujiwara et al. 2006), radar observations (e.g., Ostro et al. 2000), and occultations of background stars due to asteroids (e.g., Satō et al. 2000), but the number of asteroidal shapes obtained from these methods is much smaller than those obtained from light curves. Figure 1.3 shows the ratios of the minor to major axis lengths of asteroids obtained from Database of Asteroid Models from Inversion Technique, and we notice that about half of asteroids have the irregular shapes with the axis ratios less than 0.6.

There are several groups of asteroids that have similar proper orbital elements. Here, the proper orbital elements mean that quasi-invariants of motions of asteroids that are independent of short-term perturbations due to the gravity of the gas giants (e.g., Knezevic et al. 2002). Groups of asteroids with similar proper orbital elements are called asteroid families. Japanese astronomer Kiyotsugu Hirayama firstly identified several asteroid families (Hirayama 1918), and thus asteroid families are also called Hirayama families. Asteroids in an asteroid family usually have similar spectral properties (e.g., Zappalá et al. 1995), so that asteroids in a family are considered to be the fragments produced through collisional destruction of a parent body (e.g., Farinella et al. 1996).

There are several mechanisms that vary the proper orbital elements. One of them is the Yarkovsky effect (e.g., Farinella et al. 1998). The Yarkovsky effect mainly changes the orbital semi-major axis due to the torque induced by irradiation and rotations of asteroids. Rotations of asteroids make differences between the directions of absorption of irradiation from the sun and the directions of re-emission of radiation, which exerts the torque on asteroids and causes the semi-major axis change. The Yarkovsky effect increases the semi-major axis of prograde asteroids and decreases that of retrograde asteroids. The Yarkovsky effect significantly changes the

[2]http://astro.troja.mff.cuni.cz/projects/asteroids3D/web.php.

semi-major axis of asteroids smaller than 10 km with timescales shorter than the age of the solar system (Morbidelli and Vokrouhlický 2003). The Yarkovsky effect dissipates asteroids families for a very long period, and we can utilize this effect to measure ages of asteroid families (Spoto et al. 2015). The semi-major axis change due to the Yarkovsky effect is faster for smaller asteroids Thus if we plot diameters and proper semi-major axes of asteroids in a family on a graph that has the horizontal axis of the proper semi-major axis and the vertical axis of the inverse of the diameter, we can achieve "V-shape" occupied by the plots of asteroids. Opening angles of the V-shapes are larger for older families, and thus ages of asteroid families are estimated from the opening angles of the V-shapes. The most families have the estimated ages younger than 1 Gyr, which is much shorter than the age of the solar system.[3] Thus, asteroid families were mainly formed through recent asteroidal collisions.

Note that not all of asteroids were destructed by recent impacts but especially large asteroids may have primordial origins. O'Brien and Greenberg (2005) estimate the collisional lifetime of asteroids (or the timescale of the occurrence of destructive impacts) in the present main belt. They show that asteroids with the diameters larger than 100 km have the collisional lifetimes of about ten times longer than the age of the solar system. Thus these larger asteroids may be "fossils" of primordial asteroids (Bottke et al. 2005). O'Brien and Greenberg (2005) also show that asteroids larger than 10 km have the collisional lifetimes slightly longer than the age of the solar system, which means that some fraction of these asteroids may also be primordial.

1.2 Solar System Formation Scenario

Standard solar system formation scenarios are based on the model proposed by Safronov (1969) or the so-called Kyoto model proposed by Hayashi et al. (1985). Figure 1.4 shows the schematic picture of the solar system formation scenario based on the Kyoto model. Planets are formed in a protoplanetary disk, which is composed of gas and solid dust particles. Gas is mainly composed of hydrogen and helium. The typical size of dust particles is considered to be sub micron based on the typical size of interstellar dust particles (Mathis et al. 1977; Draine and Lee 1984). The existence of protoplanetary disks is already proved by many observations (e.g., Kitamura et al. 2002; Kwon et al. 2011), and nowadays Atacama Large Millimeter/submillimeter Array clarified even spatial structures of surface density of dust component for several protoplanetary disks (ALMA Partnership et al. 2015).

Dust particles grow up to be planetesimals, which are bodies with the typical size of 1–100 km, through collisional sticking of dust particles due to the van der Waals force. There are some difficulties for dust particles to grow to be planetesimals through collisional coalescence, although we already found some mechanisms to avoid the difficulties. One of the difficulties is radial drift problem. Gas component moves around the sun with sub-Keplerian velocity because of radial gas-pressure gradient, and dust

[3]https://newton.spacedys.com/astdys2/fam_ages/index.html.

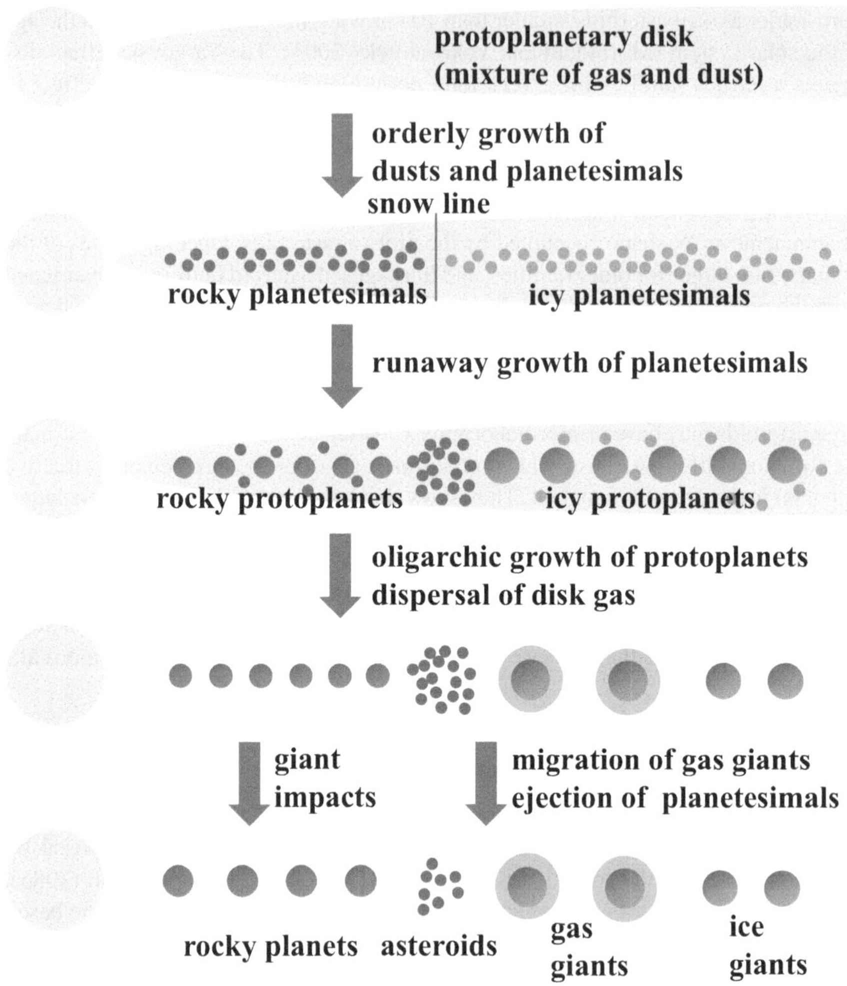

Fig. 1.4 Schematic figure that shows the formation and evolution of the solar system

particles experience head wind and loose the angular momentum. If dust particles grow with keeping compact density, the timescale for the radial drift is shorter than the growth timescale when the size of compact dust reaches about 1 m (Weidenschilling 1977), which means that dust particles fall down to the central star before they grow to be planetesimals. However, collisional sticking of dust particles produces porous aggregates (e.g., Suyama et al. 2008), and the growth timescale of porous aggregates can be shorter than that of the radial drift (Okuzumi et al. 2012). Sufficiently grown dust aggregates are compressed through gas pressure and self-gravity, and then planetesimals with the size ~1 km are formed (Kataoka et al. 2013). The radial drift problem is also overcome by steaming instability (Johansen et al. 2014).

The streaming instability produces overdense filaments composed of dust particles, and gravitational fragmentation of overdense regions directly produces planetesimals in a short timescale. The size of planetesimals produced through the streaming instability is estimated to be $\sim$100 km.

Planetesimals grow through collisional coalescence of planetesimals. They coalesce with each other owing to "glue" of the gravity between planetesimals. There are two types of the growth mode of planetesimals. If the relative velocity between planetesimals is smaller than the mutual escape velocity of planetesimals, the gravity is effective and collisional cross section is larger for larger planetesimals (gravitational focusing). This means that larger planetesimals grow faster than smaller ones, and smaller planetesimals are left behind the growth. This growth mode is called runaway growth (e.g., Wetherill and Stewart 1989; Kokubo and Ida 1996). In contrast, if the relative velocity is larger than the escape velocity, the gravity is not effective. In this case, the difference of the masses of two planetesimals decreases through the growth, and planetesimals with different masses grow with the similar growth rate. This growth mode is called orderly growth (e.g., Weidenschilling et al. 1997). In the orderly growth mode, the shape of the mass distribution of planetesimals merely shifts to massive side (Kobayashi et al. 2016).

The relative velocity between planetesimals mainly increases due to density fluctuation caused by turbulence (Ida et al. 2008; Okuzumi and Ormel 2013; Ormel and Okuzumi 2013), while the relative velocity decreases due to the collisional damping (e.g., Ohtsuki 1992). The equilibrium relative velocity between planetesimals smaller than 100 km is larger than the escape velocity from these planetesimals, and thus the growth mode for smaller planetesimals is the orderly growth (Kobayashi et al. 2016). Note that the relative velocity is surely larger than the escape velocity from the planetesimals smaller than 100 km but is still less than about ten times of the escape velocity, which does not result in the collisional destruction and results in the growth.

On the other hand, the equilibrium relative velocity between planetesimals larger than about 100 km becomes smaller than the escape velocity (Kobayashi et al. 2016). Thus the growth mode of planetesimals that reach the size of about 100 km changes to the runaway growth. Several larger planetesimals, or protoplanets, grow much faster than planetesimals smaller than 100 km, and the runaway growth produces bimodal mass distribution that is composed of several protoplanets and remnant planetesimals with radius $\sim$100 km (Kobayashi and Tanaka 2018). The mass distribution of remnant planetesimals has the peak around the radius of about 100 km, which probably remains the same until today and produces the peak around 100 km of the mass distribution of asteroids in the main belt (Fig. 1.2 and Kobayashi and Tanaka 2018). Therefore, asteroids are considered to be remnant planetesimals and they are essentially the same objects. Thus, in the explanation of collisional simulations shown after Chap. 2, we only use the term of asteroid and do not use the term of planetesimal to avoid confusions.

The runaway growth does not continue eternally. Sufficiently grown protoplanets gravitationally scatter surrounding planetesimals and increase the relative velocity between the protoplanets and planetesimals. Gravitational focusing is not effective for larger relative velocity, and thus the heat up by the protoplanets decreases the growth rate of the protoplanets themselves. Eventually, several protoplanets with the similar masses and sufficient separations are formed, and they grow through coalescence of surrounding planetesimals. This growth mode is called oligarchic growth (Kokubo and Ida 1998, 2000). The separations between protoplanets are typically about 10 Hill radii (Kokubo and Ida 1995).

Inner protoplanetary disk is hotter due to the irradiation from the sun, while outer protoplanetary disk is colder. There is the orbital radius where disk temperature is the same as the water condensation temperature. This radius is called the snow line. Solid water, or ice, exists outside of the snow line, and thus the amount of solid components outside of the snow line is larger than that inside of the snow line (Hayashi 1981), which leads to the rapid growth of protoplanets outside of the snow line. If the mass of protoplanets exceeds about 10 Earth masses, protoplanets rapidly accumulate surrounding disk gas in a runaway manner (Mizuno et al. 1978; Mizuno 1980; Stevenson 1982), which leads to the formation of gas giants. The protoplanets that eventually become Jupiter and Saturn rapidly grow because of the existence of icy component outside of the snow line and become heavier than 10 Earth masses before the dispersal of gas disk.

Disk gas damps the eccentricity of protoplanets, and protoplanets move around the sun with almost circular orbits and sufficient separations. However, after disk gas dissipates, the eccentricity increases due to the mutual gravitational perturbations between protoplanets, which leads to the orbital crossing of protoplanets and giant impacts (e.g., Wetherill 1985; Kokubo and Ida 1998). The final stage of the terrestrial planet formation is considered to be the giant impact stage. Giant impacts determine the characteristics of the present terrestrial planets, such as the number or the orbital elements of the planets (e.g., Agnor et al. 1999; Kokubo and Genda 2010). The moon is also considered to be formed by a giant impact (e.g., Ida et al. 1997; Canup 2004).

The total mass of asteroids in the present main belt is much smaller than that of solid component of the protoplanetary disk (e.g., Petit et al. 2001; Nagasawa et al. 2002). Substantial fraction of asteroids are ejected from the solar system due to, for example, the migration of the gas giants (Walsh et al. 2011), which means that impacts between planetesimals in the primordial environment are considered to be more frequent than those between asteroids in the present environment. Thus collisional deformation of planetesimals larger than 100 km probably occurs in the primordial environment although collisional destruction of 100 km-sized asteroids in the present solar system rarely occurs (O'Brien and Greenberg 2005).

1.3 Planetesimal or Asteroidal Collisions

As we introduced in Sect. 1.2, impacts between various objects including dust particles, planetesimals, and protoplanets play the essential role for the formation of the solar system, that is to say, the impact is the most important physical process for the planet formation. Collisional merging of these objects leads to the growth of objects. However, impacts do not always result in merging but also result in destruction or other types of collisions depending on mainly impact velocities and also impact angles and mass ratios of two impacting bodies. Therefore, the detailed understanding of impact outcomes is very important to discuss details of the solar system formation scenario. Here, we especially focus on planetesimal or asteroidal impacts.

Numerical simulation is one of the most powerful tools to investigate results of impacts, since the scale of planetesimals or asteroids (larger than km) is too large to be treated in laboratory experiments. Moreover, the self gravity of bodies plays important role, which is also difficult to be investigated by room experiments. Certainly, laboratory experiments of impacts for m-sized targets (e.g., Fujiwara et al. 1989; Kato et al. 1995) are applicable not only to impacts between m-sized asteroids but also to those between much larger asteroids through scaling laws (e.g., Housen and Holsapple 1990; Mizutani et al. 1990), but direct comparison between laboratory-scale-impacts and asteroid-scale-impacts is still difficult. Moreover, the recent improvement of computational power through the development of supercomputers such as K computer enhances the value of numerical simulations.

There are two major frameworks of simulation methods for impacts. One of them is shock physics codes, or hydrocodes, and the other one is particle based codes, or discrete element methods. In shock physics codes, we solve continuum equations for hydrodynamics or elastic dynamics and calculate the time evolution of field variables such as the velocity or density. Some studies include models for realistic rocky materials (e.g., Benz and Asphaug 1995; Jutzi 2015), and thus shock physics codes can treat the impacts between rocky bodies. Shock physics codes are roughly classified to particle methods and mesh methods depending on how to discretize continuum bodies. Particle methods, which include Smoothed Particle Hydrodynamics (SPH) method, utilize Lagrangian particles that mimic bulk of continuum bodies and move with the motion of continuum bodies (e.g., Jutzi and Asphaug 2015), while mesh methods utilize Eulerian meshes that are usually fixed to computational domain (e.g., Kraus et al. 2011). Impact phenomena accompany with large deformation, so that particle methods are more preferable than mesh methods. On the other hand, in particle based codes, we represent bodies as collection of particles with finite radii and rigid boundaries, and we directly assume interactions between particles. Particle based codes are also roughly classified to hard sphere models and soft sphere models. In hard sphere models, we assume the coefficient of restitution and only solve repulsion between particles (e.g., Richardson et al. 2000). In soft sphere models, we take into account finite interaction times and interactions are governed by elastic forces and frictional dissipation (e.g., Schwartz et al. 2018).

Both shock physics codes and particle based codes have merits and demerits. Shock physics codes can solve shock waves, so that they can treat hypervelocity impacts. However, time steps in shock physics codes are determined by the Courant condition and are much smaller than the timescale of gravitational reaccumulation especially for the impacts with bodies smaller than 100 km, so that it is difficult for shock physics codes to complete gravitational reaccumulation phases. On the other hand, particle based codes do not have the strict restriction for time steps especially for hard sphere models, so that the simulations of gravitational reaccumulation phases are relatively easy. However, they can treat neither shock waves nor heats generated by shock waves. Recently, some studies conduct impact simulations with hybrid codes that utilize both shock physics codes and particle based codes; they calculate initial fragmentation phases using shock physics codes and subsequent gravitational reaccumulation phases using particle based codes (e.g., Michel et al. 2001).

Various impacts are investigated in various researches so far (e.g., Michel et al. 2011; Genda et al. 2012; Ballouz et al. 2014). In this section, we introduce impact simulations in previous works in terms of the classification of impact outcomes based on Benz and Asphaug (1999) and Leinhardt and Stewart (2012) and asteroidal shapes formed through impacts based on Leinhardt et al. (2000), Jutzi and Asphaug (2015), Jutzi and Benz (2017), Schwartz et al. (2018), and Leleu et al. (2018).

1.3.1 Types of Collisional Outcomes

One of the most important measures of impact results is the mass of the largest remnants. If the mass of the largest remnants are larger than target bodies, impacts result in merging and bodies grow. Here, a target body means the larger body among two impacting objects, and an impactor body means the smaller one. Generally speaking, impacts with larger kinetic energy result in more destructive impacts. However, the degree of destruction also depends on the mass of target bodies. The larger target bodies are, the lesser the degree of destruction becomes with the same impact kinetic energy. Thus, the following total kinetic energy per unit mass Q_R is a possible good measure of the degree of destruction:

$$Q_R = \frac{(1/2)\mu_{\mathrm{imp}} v_{\mathrm{imp}}^2}{M_{\mathrm{tot}}}, \quad (1.1)$$

where v_{imp} is the impact velocity, $\mu_{\mathrm{imp}} = M_i M_{\mathrm{target}}/M_{\mathrm{tot}}$ is the reduced mass, $M_{\mathrm{tot}} = M_i + M_{\mathrm{target}}$ is the total mass, and M_i and M_{target} are masses of impactor and target bodies, respectively. For large mass difference of two impacting bodies $M_i \ll M_{\mathrm{target}}$, Q_R is simply written as

$$Q = \frac{(1/2)M_i v_{\mathrm{imp}}^2}{M_{\mathrm{target}}}. \quad (1.2)$$

The specific kinetic energy Q_R when the mass of the largest remnant becomes just half of that of a target body is called the catastrophic disruption threshold Q^*_{RD}. If Q_R is much smaller than Q^*_{RD}, the impacts result in non-destructive or merging collisions. If Q_R is comparable to or larger than Q^*_{RD}, the impacts result in catastrophic destruction. Therefore, Q^*_{RD} is one of good thresholds to judge that impacts become destructive or non-destructive. Destruction of larger bodies is prevented by the gravity

(gravity-dominated regime), while destruction of smaller bodies is prevented by material strength (strength-dominated regime).

Benz and Asphaug (1999) derived the size dependence of Q^*_D, which is the catastrophic disruption threshold for Q in Eq. (1.2), through SPH impact simulations with the material strength of intact rocks. They conducted impact simulations with various impact angles, mass ratios of two impacting bodies, and radii of target bodies but with fixed impact velocities of 3 or 5 km/s, which are the typical impact velocities in the main belt, and they derived the impact-angle-averaged Q^*_D for various target radii, which is expressed as

$$\bar{Q}^*_D = Q_0\left(\frac{R_t}{1\,\mathrm{cm}}\right)^{a_{\mathrm{pow}}} + B_0\rho_0\left(\frac{R_t}{1\,\mathrm{cm}}\right)^{b_{\mathrm{pow}}}, \tag{1.3}$$

where R_t is the radii of targets, ρ_0 is the uncompressed density of material of impacting bodies, and Q_0, B_0, a_{pow}, and b_{pow} are parameters depending on the impact velocity and target material. The detailed parameters are listed in Benz and Asphaug (1999), and the two exponents have the values of $a_{\mathrm{pow}} \approx -0.4$ and $b_{\mathrm{pow}} \approx 1.3$. Thus there is a radius that has minimum $\bar{Q}^*_D$ and this radius represents the transition between strength- and gravity-dominated regimes. They found that the transition diameter of target bodies is about 300 m. Note that Q^*_D depends on the friction coefficient of completely damaged rock (Jutzi 2015) and also the resolution of numerical simulations (Genda et al. 2015). Q^*_D is used in theoretical works to assess how impacts are destructive (e.g., Bottke et al. 2005; Kobayashi and Tanaka 2018).

The comparison between Q^*_{RD} and Q_R roughly determines whether impacts result in destructive or non-destructive ones. However, there are other types of impacts such as hit-and-run collisions (e.g., Genda et al. 2012). Moreover, Q^*_{RD} depends not only on size of target bodies as in Eq. (1.3) but also on impact angles or mass ratios of two impacting bodies. Leinhardt and Stewart (2012) extended Q^*_{RD} in gravity-dominated regime to represent the dependence on impact angles and mass ratios as follows:

$$\begin{aligned}
Q^*_{RD} &= \left(\frac{\mu_{\mathrm{imp}}}{\mu_{\mathrm{imp},\alpha}}\right)^{(3\bar{\mu}_{\mathrm{mat}}/2-2)}\left(\frac{1}{4}\frac{(q+1)^2}{q}\right)^{(2/(3\bar{\mu}_{\mathrm{mat}})-1)} Q^*_{RD,q=1,\theta_{\mathrm{imp}}=0},\\
q &= M_i/M_{\mathrm{target}},\\
\mu_{\mathrm{imp},\alpha} &= \frac{\alpha_{\mathrm{imp}}M_iM_{\mathrm{target}}}{\alpha_{\mathrm{imp}}M_i + M_{\mathrm{target}}},\\
\alpha_{\mathrm{imp}} &= \frac{3R_il^2 - l^3}{4R_i^3},\\
l &= \begin{cases} (R_i + R_t)(1-\sin\theta_{\mathrm{imp}}) & R_t < R_i + (R_i+R_t)\sin\theta_{\mathrm{imp}} \\ 2R_i & R_t > R_i + (R_i+R_t)\sin\theta_{\mathrm{imp}} \end{cases},
\end{aligned} \tag{1.4}$$

where R_i is the radii of impactors, θ_{imp} is the impact angle, $Q^*_{RD,q=1,\theta_{\mathrm{imp}}=0}$ is the catastrophic disruption threshold for equal-mass and head-on collisions, and $\bar{\mu}_{\mathrm{mat}}$ is the material dependent scaling parameter with $1/3 \leq \bar{\mu}_{\mathrm{mat}} \leq 2/3$. Leinhardt and

Stewart (2012) determined $Q^*_{RD,q=1,\theta_{\rm imp}=0}$ using scaling laws, but it can also be determined by numerical simulations as done in Benz and Asphaug (1999). For detailed classification of impact types, they also set the following critical impact angle (see also Asphaug 2010):

$$\theta_{\rm imp,crit} = \sin^{-1}\left(\frac{R_t}{R_t + R_i}\right). \tag{1.5}$$

Using Q^*_{RD}, $\theta_{\rm imp,crit}$, and two-body escape velocity $v_{\rm esc}$, we can roughly classify impact types as follows: merging collisions for $v_{\rm imp} \leq v_{\rm esc}$, partial accretion for $\theta_{\rm imp} \leq \theta_{\rm imp,crit}$ and $Q_{RD} < Q^*_{RD}$, hit-and-run collisions for $\theta_{\rm imp} > \theta_{\rm imp,crit}$ and $Q_{RD} < Q^*_{RD}$, and catastrophic destruction for $Q_{RD} \geq Q^*_{RD}$.

1.3.2 Collisions as the Origin of Asteroidal Shapes

In Sect. 1.3.1, we focused on the types of impacts and mass of the largest remnants. There are many other interesting features for results of impacts such as the rotation of impact outcomes. In this dissertation, we especially focus on shapes of impact outcomes, because shapes of impact outcomes are sensitive to impact conditions (e.g., Jutzi and Asphaug 2015) and we already know shapes of many actual asteroids (see Fig. 1.3).

Shapes of asteroids are mainly determined by asteroidal collisions, although there are other mechanisms to deform asteroidal shapes such as spin-up due to the Yarkovsky-O'Keefe-Radzievskii-Paddack (YORP) effect (e.g., Richardson et al. 2005) or tidal disruption due to planets (e.g., Walsh and Richardson 2006). The YORP effect is similar to the Yarkovsky effect and is also caused by irradiation, but the YORP effect changes the rotation rate of asteroids (Rubincam 2000). If rotation rate becomes too large due to the YORP effect, centrifugal forces due to the rotation exceeds the self gravity and shapes are deformed (e.g., Pravec et al. 2002). However, the timescale of rotation-rate changes due to the YORP effect for asteroids larger than 10 km is much longer than the age of the solar system (Rubincam 2000), and thus the YORP deformation is not important for the larger asteroids. The deformation due to tidal disruption is also inefficient, because tidal disruption of asteroids only occurs when asteroids closely encounter with large planets although there are no large planets in the main belt.

Shapes of impact outcomes are investigated in several previous researches using the particle based method (Leinhardt et al. 2000), the SPH method (Jutzi and Asphaug 2015; Jutzi and Benz 2017; Leleu et al. 2018), and the hybrid method (Schwartz et al. 2018). Here, we briefly introduce each study.

Leinhardt et al. (2000) conducted impact simulations between km-sized rubble pile bodies with impact velocities smaller than 10 m/s, which are comparable to the two-body escape velocity of the rubble pile bodies. They investigated the dependence of impact outcomes on impact velocities, impact angles, mass ratios of two impacting bodies, and even rotation rates of impacting bodies. They found that equal-mass impacts can form elongated or bilobed shapes, while all shapes formed through unequal-mass impacts are spherical. Although they investigated the detailed dependence of the shapes of impact outcomes on the impact velocities and angles, they conducted simulations with two values of the mass ratios (1/1 and 1/8) and they do not investigate impact outcomes formed through destructive impacts.

Jutzi and Asphaug (2015), Jutzi and Benz (2017), and Schwartz et al. (2018) focus on how to form the bilobed shape of Comet 67P/Churyumov-Gerasimenko, which has the size of 1 km. Jutzi and Asphaug (2015) investigated similar-mass and low-velocity ($\approx$1 m/s) impacts, and showed that bilobed shapes are formed through impacts with moderate initial angular momentum. However, impacts with the impact velocity $\approx$1 m/s occurs only in the primordial environment, and it is quite unlikely that such shapes of km-sized bodies survive until today without experiencing destructive collisions. Jutzi and Benz (2017) showed that sub-catastrophic destruction of rotating elongated body with the impact velocity of $\approx$300 m/s forms a bilobed shape through the fission of a rotating elongated body into two remnants and reaccumulation of the remnants. Schwartz et al. (2018) showed that bilobed shapes are also formed through the coalescence of two spherical remnants during the reaccumulation phase after catastrophic destruction.

Leleu et al. (2018) investigated the production procedure of the peculiar shapes of the inner and small Saturnian moons. Among six inner Saturnian moons, there are three irregularly shaped moons, that is, Pan and Atlas have the flat shapes and Prometheus has the elongated shape. Moons produced from the rings of Saturn have similar masses and small orbital eccentricities due to the oblateness of Saturn, which leads to similar-mass and low-velocity impacts between Saturnian moons. They investigated the probability distribution of impact conditions around Saturn and shapes formed through impacts with probable impact conditions, and they found that similar-mass and low-velocity impacts around Saturn can explain the current fraction of the irregular shapes of the small Saturnian moons.

Although there are several previous researches investigating shapes formed through impacts, almost all of them focus on the shapes of specific objects and do not investigated comprehensive shapes formed through various impacts. They also do not investigate shapes quantitatively, and then we cannot statistically compare the results of simulations with the shape distribution of actual asteroids. Thus, in this dissertation, we conduct various impact simulations and investigate shapes of impact outcomes quantitatively in terms of axis ratios.

References

C.B. Agnor, R.M. Canup, H.F. Levison, On the character and consequences of large impacts in the late stage of terrestrial planet formation. Icarus **142**, 219–237 (1999). https://doi.org/10.1006/icar.1999.6201

ALMA Partnership, C.L. Brogan, L.M. Pérez, T.R. Hunter, W.R.F. Dent, A.S. Hales, R.E. Hills, S. Corder, E.B. Fomalont, C. Vlahakis, Y. Asaki, D. Barkats, A. Hirota, J.A. Hodge, C.M.V. Impellizzeri, R. Kneissl, E. Liuzzo, R. Lucas, N. Marcelino, S. Matsushita, K. Nakanishi, N. Phillips, A.M.S. Richards, I. Toledo, R. Aladro, D. Broguiere, J.R. Cortes, P.C. Cortes, D. Espada, F. Galarza, D. Garcia-Appadoo, L. Guzman-Ramirez, E.M. Humphreys, T. Jung, S. Kameno, R.A. Laing, S. Leon, G. Marconi, A. Mignano, B. Nikolic, L.-A. Nyman, M. Radiszcz, A. Remijan, J.A. Rodón, T. Sawada, S. Takahashi, R.P.J. Tilanus, B. Vila Vilaro, L.C. Watson, T. Wiklind, E. Akiyama, E. Chapillon, I. de Gregorio-Monsalvo, J. Di Francesco, F. Gueth, A. Kawamura, C.-F. Lee, Q. Nguyen Luong, J. Mangum, V. Pietu, P. Sanhueza, K. Saigo, S. Takakuwa, C. Ubach, T. van Kempen, A. Wootten, A. Castro-Carrizo, H. Francke, J. Gallardo, J. Garcia, S. Gonzalez, T. Hill, T. Kaminski, Y. Kurono, H.-Y. Liu, C. Lopez, F. Morales, K. Plarre, G. Schieven, L. Testi, L. Videla, E. Villard, P. Andreani, J.E. Hibbard, K. Tatematsu, The 2014 ALMA long baseline campaign: first results from high angular resolution observations toward the HL Tau region. Astrophys. J. **808**, L3 (2015). https://doi.org/10.1088/2041-8205/808/1/L3

E. Asphaug, Similar-sized collisions and the diversity of planets. Chem. Erde. Geochem. **70**, 199–219 (2010). https://doi.org/10.1016/j.chemer.2010.01.004

R.-L. Ballouz, D.C. Richardson, P. Michel, S.R. Schwartz, Rotation-dependent catastrophic disruption of gravitational aggregates. Astrophys. J. **789**, 158 (2014). https://doi.org/10.1088/0004-637X/789/2/158

W. Benz, E. Asphaug, Simulations of brittle solids using smooth particle hydrodynamics. Comput. Phys. Comm. **87**, 253–265 (1995). https://doi.org/10.1016/0010-4655(94)00176-3

W. Benz, E. Asphaug, Catastrophic disruptions revisited. Icarus **142**, 5–20 (1999). https://doi.org/10.1006/icar.1999.6204

R.P. Binzel, S. Xu, S.J. Bus, E. Bowell, Origins for the near-earth asteroids. Science **257**(5071), 779–782 (1992). https://doi.org/10.1126/science.257.5071.779

W.F. Bottke, D.D. Durda, D. Nesvorný, R. Jedicke, A. Morbidelli, D. Vokrouhlický, H. Levison, The fossilized size distribution of the main asteroid belt. Icarus **175**, 111–140 (2005). https://doi.org/10.1016/j.icarus.2004.10.026

R.M. Canup, Origin of terrestrial planets and the earth-moon system. Phys. Today **57**(4), 56–62 (2004). https://doi.org/10.1063/1.1752423

B.T. Draine, H.M. Lee, Optical properties of interstellar graphite and silicate grains. Astrophys. J. **285**, 89–108 (1984). https://doi.org/10.1086/162480

J. Ďurech, V. Sidorin, M. Kaasalainen, DAMIT: a database of asteroid models. Astron. Astrophys. **513**, A46 (2010). https://doi.org/10.1051/0004-6361/200912693

P. Farinella, D.R. Davis, F. Marzari, Asteroid families, old and young, in *Completing the Inventory of the Solar System*, ed. by T. Rettig, J.M. Hahn, vol. 107 of Astronomical Society of the Pacific Conference Series (1996), pp. 45–55

P. Farinella, D. Vokrouhlický, W.K. Hartmann, Meteorite delivery via Yarkovsky orbital drift. Icarus **132**, 378–387 (1998). https://doi.org/10.1006/icar.1997.5872

A. Fujiwara, P. Cerroni, D.R. Davis, E. Ryan, M. di Martino, K. Holsapple, K. Housen, Experiments and scaling laws for catastrophic collisions, in *Asteroids II*, ed. by R.P. Binzel, T. Gehrels, M.S. Matthews (1989), pp. 240–265

A. Fujiwara, J. Kawaguchi, D.K. Yeomans, M. Abe, T. Mukai, T. Okada, J. Saito, H. Yano, M. Yoshikawa, D.J. Scheeres, O. Barnouin-Jha, A.F. Cheng, H. Demura, R.W. Gaskell, N. Hirata, H. Ikeda, T. Kominato, H. Miyamoto, A.M. Nakamura, R. Nakamura, S. Sasaki, K. Uesugi, The rubble-pile asteroid Itokawa as observed by Hayabusa. Science **312**, 1330–1334 (2006). https://doi.org/10.1126/science.1125841

H. Genda, E. Kokubo, S. Ida, Merging criteria for giant impacts of protoplanets. Astrophys. J. **744**, 137 (2012). https://doi.org/10.1088/0004-637X/744/2/137

H. Genda, T. Fujita, H. Kobayashi, H. Tanaka, Y. Abe, Resolution dependence of disruptive collisions between planetesimals in the gravity regime. Icarus **262**, 58–66 (2015). https://doi.org/10.1016/j.icarus.2015.08.029

C. Hayashi, Structure of the solar nebula, growth and decay of magnetic fields and effects of magnetic and turbulent viscosities on the nebula. Prog. Theor. Phys. Suppl. **70**, 35–53 (1981). https://doi.org/10.1143/PTPS.70.35

C. Hayashi, K. Nakazawa, Y. Nakagawa, Formation of the solar system, in *Protostars and Planets II*, ed. by D.C. Black, M.S. Matthews (1985), pp. 1100–1153

K. Hirayama, Groups of asteroids probably of common origin. Astron. J. **31**, 185–188 (1918). https://doi.org/10.1086/104299

K.R. Housen, K.A. Holsapple, On the fragmentation of asteroids and planetary satellites. Icarus **84**, 226–253 (1990). https://doi.org/10.1016/0019-1035(90)90168-9

D.W. Hughes, G.H.A. Cole, The asteroidal sphericity limit. Mon. Not. R. Astron. Soc. **277**, 99–105 (1995). https://doi.org/10.1093/mnras/277.1.99

S. Ida, R.M. Canup, G.R. Stewart, Lunar accretion from an impact-generated disk. Nature **389**, 353–357 (1997). https://doi.org/10.1038/38669

S. Ida, T. Guillot, A. Morbidelli, Accretion and destruction of planetesimals in turbulent disks. Astrophys. J. **686**, 1292–1301 (2008). https://doi.org/10.1086/591903

A. Johansen, J. Blum, H. Tanaka, C. Ormel, M. Bizzarro, H. Rickman, The multifaceted planetesimal formation process, in *Protostars and Planets VI* (2014), pp. 547–570

M. Jutzi, SPH calculations of asteroid disruptions: the role of pressure dependent failure models. Planet. Space Sci. **107**, 3–9 (2015). https://doi.org/10.1016/j.pss.2014.09.012

M. Jutzi, E. Asphaug, The shape and structure of cometary nuclei as a result of low-velocity accretion. Science **348**, 1355–1358 (2015). https://doi.org/10.1126/science.aaa4747

M. Jutzi, W. Benz, Formation of bi-lobed shapes by sub-catastrophic collisions. A late origin of comet 67P's structure. Astron. Astrophys. **597**, A62 (2017). https://doi.org/10.1051/0004-6361/201628964

M. Kaasalainen, J. Torppa, Optimization methods for asteroid lightcurve inversion. I. Shape determination. Icarus **153**, 24–36 (2001). https://doi.org/10.1006/icar.2001.6673

M. Kaasalainen, J. Torppa, K. Muinonen, Optimization methods for asteroid lightcurve inversion. II. The complete inverse problem. Icarus **153**, 37–51 (2001). https://doi.org/10.1006/icar.2001.6674

A. Kataoka, H. Tanaka, S. Okuzumi, K. Wada, Fluffy dust forms icy planetesimals by static compression. Astron. Astrophys. **557**, L4 (2013). https://doi.org/10.1051/0004-6361/201322151

M. Kato, Y.-I. Iijima, M. Arakawa, Y. Okimura, A. Fujimura, N. Maeno, H. Mizutani, Ice-on-ice impact experiments. Icarus **113**(2), 423–441 (1995). https://doi.org/10.1006/icar.1995.1032

Y. Kitamura, M. Momose, S. Yokogawa, R. Kawabe, M. Tamura, S. Ida, Investigation of the physical properties of protoplanetary disks around T Tauri stars by a 1 arcsecond imaging survey: evolution and diversity of the disks in their accretion stage. Astrophys. J. **581**, 357–380 (2002). https://doi.org/10.1086/344223

Z. Knezevic, A. Lemaître, A. Milani, The determination of asteroid proper elements (2002), pp. 603–612

H. Kobayashi, H. Tanaka, From planetesimal to planet in turbulent disks. II. Formation of gas giant planets. Astrophys. J. **862**, 127 (2018). https://doi.org/10.3847/1538-4357/aacdf5

H. Kobayashi, H. Tanaka, S. Okuzumi, From planetesimals to planets in turbulent protoplanetary disks. I. Onset of runaway growth. Astrophys. J. **817**, 105 (2016). https://doi.org/10.3847/0004-637X/817/2/105

E. Kokubo, H. Genda, Formation of terrestrial planets from protoplanets under a realistic accretion condition. Astrophys. J. **714**, L21–L25 (2010). https://doi.org/10.1088/2041-8205/714/1/L21

E. Kokubo, S. Ida, Orbital evolution of protoplanets embedded in a swarm of planetesimals. Icarus **114**, 247–257 (1995). https://doi.org/10.1006/icar.1995.1059

E. Kokubo, S. Ida, On runaway growth of planetesimals. Icarus **123**, 180–191 (1996). https://doi.org/10.1006/icar.1996.0148

E. Kokubo, S. Ida, Oligarchic growth of protoplanets. Icarus **131**, 171–178 (1998). https://doi.org/10.1006/icar.1997.5840

E. Kokubo, S. Ida, Formation of protoplanets from planetesimals in the solar nebula. Icarus **143**, 15–27 (2000). https://doi.org/10.1006/icar.1999.6237

R.G. Kraus, L.E. Senft, S.T. Stewart, Impacts onto H_2O ice: scaling laws for melting, vaporization, excavation, and final crater size. Icarus **214**, 724–738 (2011). https://doi.org/10.1016/j.icarus.2011.05.016

W. Kwon, L.W. Looney, L.G. Mundy, Resolving the circumstellar disk of HL Tauri at millimeter wavelengths. Astrophys. J. **741**, 3 (2011). https://doi.org/10.1088/0004-637X/741/1/3

Z.M. Leinhardt, S.T. Stewart, Collisions between Gravity-dominated Bodies. I. Outcome regimes and scaling laws. Astrophys. J. **745**, 79 (2012). https://doi.org/10.1088/0004-637X/745/1/79

Z.M. Leinhardt, D.C. Richardson, T. Quinn, Direct N-body simulations of rubble pile collisions. Icarus **146**, 133–151 (2000). https://doi.org/10.1006/icar.2000.6370

A. Leleu, M. Jutzi, M. Rubin, The peculiar shapes of Saturn's small inner moons as evidence of mergers of similar-sized moonlets. Nature Astron. **2**, 555–561 (2018). https://doi.org/10.1038/s41550-018-0471-7

C.H. Lineweaver, M. Norman, The potato radius: a lower minimum size for dwarf planets. ArXiv e-prints (2010)

J.S. Mathis, W. Rumpl, K.H. Nordsieck, The size distribution of interstellar grains. Astrophys. J. **217**, 425–433 (1977). https://doi.org/10.1086/155591

P. Michel, W. Benz, P. Tanga, D.C. Richardson, Collisions and gravitational reaccumulation: forming asteroid families and satellites. Science **294**, 1696–1700 (2001). https://doi.org/10.1126/science.1065189

P. Michel, M. Jutzi, D.C. Richardson, W. Benz, The asteroid Veritas: an intruder in a family named after it? Icarus **211**, 535–545 (2011). https://doi.org/10.1016/j.icarus.2010.10.012

P. Michel, A. Morbidelli, W.F. Bottke, Origin and dynamics of near earth objects. Comptes Rendus Physique **6**(3), 291–301 (2005). https://doi.org/10.1016/j.crhy.2004.12.013. The near earth objects: possible impactors of the earth

H. Mizuno, Formation of the giant planets. Prog. Theor. Phys. **64**, 544–557 (1980). https://doi.org/10.1143/PTP.64.544

H. Mizuno, K. Nakazawa, C. Hayashi, Instability of a gaseous envelope surrounding a planetary core and formation of giant planets. Prog. Theor. Phys. **60**, 699–710 (1978). https://doi.org/10.1143/PTP.60.699

H. Mizutani, Y. Takagi, S.-I. Kawakami, New scaling laws on impact fragmentation. Icarus **87**, 307–326 (1990). https://doi.org/10.1016/0019-1035(90)90136-W

A. Morbidelli, D. Vokrouhlický, The Yarkovsky-driven origin of near-earth asteroids. Icarus **163**, 120–134 (2003). https://doi.org/10.1016/S0019-1035(03)00047-2

M. Nagasawa, S. Ida, H. Tanaka, Excitation of orbital inclinations of asteroids during depletion of a protoplanetary disk: dependence on the disk configuration. Icarus **159**, 322–327 (2002). https://doi.org/10.1006/icar.2002.6925

D.P. O'Brien, R. Greenberg, The collisional and dynamical evolution of the main-belt and NEA size distributions. Icarus **178**, 179–212 (2005). https://doi.org/10.1016/j.icarus.2005.04.001

K. Ohtsuki, Evolution of random velocities of planetesimals in the course of accretion. Icarus **98**, 20–27 (1992). https://doi.org/10.1016/0019-1035(92)90202-I

S. Okuzumi, C.W. Ormel, The fate of planetesimals in turbulent disks with dead zones. I. The turbulent stirring recipe. Astrophys. J. **771**, 43 (2013). https://doi.org/10.1088/0004-637X/771/1/43

S. Okuzumi, H. Tanaka, H. Kobayashi, K. Wada, Rapid coagulation of porous dust aggregates outside the snow line: a pathway to successful icy planetesimal formation. Astrophys. J. **752**, 106 (2012). https://doi.org/10.1088/0004-637X/752/2/106

C.W. Ormel, S. Okuzumi, The fate of planetesimals in turbulent disks with dead zones. II. Limits on the viability of runaway accretion. Astrophys. J. **771**, 44 (2013). https://doi.org/10.1088/0004-637X/771/1/44

S.J. Ostro, R. Scott, Hudson, M.C. Nolan, J.-L. Margot, D.J. Scheeres, D.B. Campbell, C. Magri, J.D. Giorgini, D.K. Yeomans, Radar observations of asteroid 216 kleopatra. Science **288**(5467), 836–839 (2000). https://doi.org/10.1126/science.288.5467.836

J.-M. Petit, A. Morbidelli, J. Chambers, The primordial excitation and clearing of the asteroid belt. Icarus **153**, 338–347 (2001). https://doi.org/10.1006/icar.2001.6702

P. Pravec, A.W. Harris, T. Michalowski. Asteroid rotations 113–122 (2002)

D.C. Richardson, T. Quinn, J. Stadel, G. Lake, Direct large-scale N-body simulations of planetesimal dynamics. Icarus **143**, 45–59 (2000). https://doi.org/10.1006/icar.1999.6243

D.C. Richardson, P. Elankumaran, R.E. Sanderson, Numerical experiments with rubble piles: equilibrium shapes and spins. Icarus **173**, 349–361 (2005). https://doi.org/10.1016/j.icarus.2004.09.007

D.P. Rubincam, Radiative spin-up and spin-down of small asteroids. Icarus **148**, 2–11 (2000). https://doi.org/10.1006/icar.2000.6485

V.S. Safronov, Evoliutsiia doplanetnogo oblaka (1969)

I. Satō, L. Šarounová, H. Fukushima, Size and shape of trojan asteroid Diomedes from its occultation and photometry. Icarus **145**, 25–32 (2000). https://doi.org/10.1006/icar.1999.6316

S.R. Schwartz, P. Michel, M. Jutzi, S. Marchi, Y. Zhang, D.C. Richardson, Catastrophic disruptions as the origin of bilobate comets. Nature Astron. **2**, 379–382 (2018). https://doi.org/10.1038/s41550-018-0395-2

F. Spoto, A. Milani, Z. Kneževiá, Asteroid family ages. Icarus **257**, 275–289 (2015). https://doi.org/10.1016/j.icarus.2015.04.041

D.J. Stevenson, Formation of the giant planets. P & SS **30**, 755–764 (1982). https://doi.org/10.1016/0032-0633(82)90108-8

T. Suyama, K. Wada, H. Tanaka, Numerical simulation of density evolution of dust aggregates in protoplanetary disks. I. Head-on collisions. Astrophys. J. **684**, 1310–1322 (2008). https://doi.org/10.1086/590143

K.J. Walsh, D.C. Richardson, Binary near-earth asteroid formation: rubble pile model of tidal disruptions. Icarus **180**(1), 201–216 (2006). https://doi.org/10.1016/j.icarus.2005.08.015

K.J. Walsh, A. Morbidelli, S.N. Raymond, D.P. O'Brien, A.M. Mandell, A low mass for Mars from Jupiter's early gas-driven migration. Nature **475**, 206–209 (2011). https://doi.org/10.1038/nature10201

S.J. Weidenschilling, Aerodynamics of solid bodies in the solar nebula. Mon. Not. R. Astron. Soc. **180**, 57–70 (1977). https://doi.org/10.1093/mnras/180.1.57

S.J. Weidenschilling, D. Spaute, D.R. Davis, F. Marzari, K. Ohtsuki, Accretional evolution of a planetesimal swarm. Icarus **128**, 429–455 (1997). https://doi.org/10.1006/icar.1997.5747

G.W. Wetherill, Occurrence of giant impacts during the growth of the terrestrial planets. Science **228**, 877–879 (1985). https://doi.org/10.1126/science.228.4701.877

G.W. Wetherill, G.R. Stewart, Accumulation of a swarm of small planetesimals. Icarus **77**, 330–357 (1989). https://doi.org/10.1016/0019-1035(89)90093-6

V. Zappalá, P. Bendjoya, A. Cellino, P. Farinella, C. Froeschlé, Asteroid families: search of a 12,487-asteroid sample using two different clustering techniques. Icarus **116**(2), 291–314 (1995). https://doi.org/10.1006/icar.1995.1127

Chapter 2
Method

Abstract Numerical simulations are the most powerful tool to investigate resultant shapes of asteroids produced through various collisional events. To conduct such numerical simulations, we need knowledge about not only hydrodynamics but also elastic dynamics for intact rocks, fracturing of rocky materials, and friction of completely damaged rocks. In this chapter, we introduce the detailed methods necessary for numerical simulations of asteroidal collisions, i.e., equations for elastic dynamics, discretized equations for smoothed particle hydrodynamics method, equation of states for impact simulations, a model for fracture of rocks, a model for friction of damaged rocks, parallelization of simulation codes, and a time development procedure.

Keywords Elastic dynamics · Smoothed particle hydrodynamics method · Fracture model for rocky materials · Friction model for granular materials

2.1 Elastic Dynamics

2.1.1 Basic Equations for Elastic Dynamics

Elastic bodies mean bodies that always return to those initial shapes owing to internal stresses under any deformation. Intact asteroids, i.e., rocky material, are also elastic bodies when deformation is small. Thus our simulation method is based on the elastic dynamics. Basic equations include not only the equation of motion or the equation of continuity but also the equation to calculate the stress tensor from the strain. The SPH method is a particle method, which basically requires Lagrangian formulation, and thus we introduce equations in Lagrange-type formulation.

The Hooke's law describes the relationship between the stress and the strain for elastic bodies with small deformation. According to the Hooke's law, the stress tensor $\sigma^{\alpha\beta}$ is proportional to the strain, which is the displacement per unit length. $\sigma^{\alpha\beta}$ means the force per unit area acting on a surface of a part of elastic bodies, and β indicates

K. Sugiura, *Development of a Numerical Simulation Method for Rocky Body Impacts and Theoretical Analysis of Asteroidal Shapes*, Springer Theses,
https://doi.org/10.1007/978-981-15-3722-6_2

the direction of the force and α indicates the direction perpendicular to the surface. Here, Greek letters of α or β correspond to each direction x, y, or z. According to the Hooke's law, the stress tensor is expressed as

$$\sigma^{\alpha\beta} = 2\mu\left(\epsilon^{\alpha\beta} - \frac{1}{3}\epsilon^{\gamma\gamma}\delta^{\alpha\beta}\right) + K\epsilon^{\gamma\gamma}\delta^{\alpha\beta}, \tag{2.1}$$

where μ is the shear modulus, K is the bulk modulus, and $\delta^{\alpha\beta}$ is the Kronecker delta. We assume a summation convention over repeated indices of Greek letters. $\epsilon^{\alpha\beta}$ is the strain tensor and written as,

$$\epsilon^{\alpha\beta} = \frac{1}{2}\left(\frac{\partial\zeta^{\alpha}}{\partial x^{\beta}} + \frac{\partial\zeta^{\beta}}{\partial x^{\alpha}}\right), \tag{2.2}$$

where $\zeta^{\alpha} = x^{\alpha} - x_0^{\alpha}$, $\boldsymbol{x}$ is the position vector, and $\boldsymbol{x}_0$ is the position vector when bodies are not deformed. The first term of Eq. (2.1) shows the deviatoric stress tensor, and we define

$$S^{\alpha\beta} \equiv 2\mu\left(\epsilon^{\alpha\beta} - \frac{1}{3}\epsilon^{\gamma\gamma}\delta^{\alpha\beta}\right). \tag{2.3}$$

The second term of Eq. (2.1) shows the isotropic stress, i.e., the pressure, and we define $P \equiv -K\epsilon^{\gamma\gamma}$. For numerical simulations, it is convenient to use the relationship between the pressure, density, and internal energy, i.e., equation of states. $\epsilon^{\gamma\gamma}$ is related to the change of volume, and the pressure with small deformations is expressed as

$$P = \frac{K}{\rho_0}(\rho - \rho_0) = C_s^2(\rho - \rho_0), \tag{2.4}$$

where ρ is the density, ρ_0 is the density when bodies are not deformed, and $C_s = \sqrt{K/\rho_0}$ is the bulk sound speed. Note that positive pressures represent repulsive forces, while negative pressures represent tensile forces. Thus the stress tensor is represented as

$$\sigma^{\alpha\beta} = -P\delta^{\alpha\beta} + S^{\alpha\beta}. \tag{2.5}$$

However, the strain tensor defined as Eq. (2.2) does not become zero for rigid-body rotations, which leads to nonzero deviatoric stress tensor calculated from Eq. (2.3). Note that no deformation occurs for rigid-body rotations and the strain tensor should be zero. To avoid this problem, we solve the time-evolution equation of the deviatoric stress tensor (e.g., Libersky and Petschek 1991). The equation is achieved by differentiating Eq. (2.3) and represented as

$$\frac{dS^{\alpha\beta}}{dt} = 2\mu\left(\dot{\epsilon}^{\alpha\beta} - \frac{1}{3}\delta^{\alpha\beta}\dot{\epsilon}^{\gamma\gamma}\right) + S^{\alpha\gamma}R^{\beta\gamma} + S^{\beta\gamma}R^{\alpha\gamma}, \tag{2.6}$$

where d/dt represents Lagrangian time derivative and $\dot{\epsilon}^{\alpha\beta}$ and $R^{\alpha\gamma}$ represent the strain rate tensor and the rotation rate tensor, respectively. The strain rate tensor and the rotation rate tensor are expressed as

$$\dot{\epsilon}^{\alpha\beta} = \frac{1}{2}\left(\frac{\partial v^\alpha}{\partial x^\beta} + \frac{\partial v^\beta}{\partial x^\alpha}\right),$$
$$R^{\alpha\beta} = \frac{1}{2}\left(\frac{\partial v^\alpha}{\partial x^\beta} - \frac{\partial v^\beta}{\partial x^\alpha}\right), \tag{2.7}$$

where $\boldsymbol{v}$ is the velocity vector. The first term of Eq. (2.6) is achieved straightforwardly by differentiating Eq (2.3). Equation (2.3) is written for the coordinate system fixed to bodies, but we utilize the laboratory coordinate system. Thus we need to consider the difference of two coordinate systems when bodies rotate. The second term of Eq. (2.6) represents this difference and is derived from the derivative of bases.

The acceleration of a part of elastic bodies is determined by the space derivative of the stress tensor. The equation of motion, which determines the acceleration of elastic bodies, is expressed as

$$\frac{dv^\alpha}{dt} = \frac{1}{\rho}\frac{\partial \sigma^{\alpha\beta}}{\partial x^\beta}. \tag{2.8}$$

The density change is determined by the divergence of the velocity field. The equation of continuity, which determines the change rate of the density, is expressed as

$$\frac{d\rho}{dt} = -\rho\frac{\partial v^\alpha}{\partial x^\alpha}. \tag{2.9}$$

We can treat the motion of elastic bodies by solving the equation of state (2.4), the time-evolution equation of the deviatoric stress tensor (2.6), the equation of motion (2.8), and the equation of continuity (2.9). However, if we want to solve the temperature change, we need to additionally solve the equation of energy. The change of the specific internal energy u due to works done by internal stresses is basically determined by the product of the stress tensor and the velocity gradient, i.e., deformation per unit time, and thus the equation of energy is expressed as,

$$\frac{du}{dt} = \frac{1}{\rho}\sigma^{\alpha\beta}\frac{\partial v^\alpha}{\partial x^\beta}. \tag{2.10}$$

2.1.2 *Basic Concept of SPH Method*

To numerically solve the Eqs. (2.6), (2.8)–(2.10), we need to discretize these equations. The SPH method is one of numerical concepts to discretize these equations. The SPH method is originally the numerical method for fluid dynamics using Lagrangian "particles" and firstly developed by Gingold and Monaghan (1977) and Lucy (1977) to solve hydrodynamics problems in astrophysics fields. Each SPH particle mimics a fluid element, and we represent fluid bodies as swarms of SPH particles. Recently, the SPH method is extended to elastic dynamics (e.g., Gray et al. 2001; Libersky and Petschek 1991; Sugiura and Inutsuka 2017), dynamics of brittle solids (Benz and Asphaug 1994, 1995), and dynamics of granular materials (Jutzi 2015).

Each SPH particle is not a point particle but has a kind of finite radius. In the framework of the SPH method, the density at the position $\boldsymbol{x}$ is represented as

$$\rho(\boldsymbol{x}) = \sum_j m_j W(|\boldsymbol{x} - \boldsymbol{x}_j|, h), \tag{2.11}$$

where m_j is the mass of the jth SPH particle, $\boldsymbol{x}_j$ is the position of the jth particle, $W(|\boldsymbol{x} - \boldsymbol{x}_j|, h)$ is a kernel function, and h is a smoothing length. Hereafter, subscripts of Roman letters mean the particle number. The kernel function $W(|\boldsymbol{x} - \boldsymbol{x}_j|, h)$ represents shape of a kind of distribution of the jth particle, and the smoothing length determines the width of the broadening of each SPH particle (h is usually set to be comparable to average particle spacing). In Eq. (2.11), $m_j W(|\boldsymbol{x} - \boldsymbol{x}_j|, h)$ represents the density distribution contributed from the jth particle, and thus the total density distribution is represented by superposition of the density distribution contributed from nearby SPH particles. Note that the kernel function has following properties: (I) the kernel function converges to a Dirac δ function when $h \to 0$; (II) the kernel function is normalized so that the spatial integration of the kernel function becomes unity; (III) the kernel function $W(r, h)$ becomes 0 for $r \to \infty$. The kernel function has various forms, and here we introduce two types of the kernel function; Gaussian and cubic spline kernels. The Gaussian kernel is expressed as

$$W(r, h) = \left[\frac{1}{h\sqrt{\pi}}\right]^{d_{\mathrm{dim}}} e^{-r^2/h^2}, \tag{2.12}$$

where d_{dim} is a spatial dimension, and the cubic spline kernel for three dimensional problems is expressed as

$$W(r, h) = \frac{1}{\pi h^3} \begin{cases} 1 - \frac{3}{2}\left(\frac{r}{h}\right)^2 + \frac{3}{4}\left(\frac{r}{h}\right)^3 & 0 \le \frac{r}{h} < 1 \\ \frac{1}{4}\left(2 - \frac{r}{h}\right)^3 & 1 \le \frac{r}{h} < 2 \\ 0 & 2 \le \frac{r}{h} \end{cases}. \tag{2.13}$$

Smoothing lengths should be comparable to local average particle spacing. Thus if the particle spacing, or the density, largely varies in space and in time, we need to adjust smoothing lengths to be local average particle spacing. However, all simulations in this dissertation do not accompany with large density variation. Thus we use constant smoothing lengths in space and in time, and we set smoothing lengths to be initial particle spacing.

A physical quantity f_i at the position of the ith particle $\boldsymbol{x}_i$ is approximated by the convolution of f at vicinity points, which is given by

$$f_i = \int f(\boldsymbol{x}) W(|\boldsymbol{x} - \boldsymbol{x}_i|, h) d\boldsymbol{x}. \tag{2.14}$$

In the same manner, the convolution of the spatial derivative of the physical quantity is given by

$$\begin{aligned} \frac{\partial f_i}{\partial x_i^\alpha} &= \int \frac{\partial f(\boldsymbol{x})}{\partial x^\alpha} W(|\boldsymbol{x} - \boldsymbol{x}_i|, h) d\boldsymbol{x} \\ &= \int f(\boldsymbol{x}) \frac{\partial}{\partial x_i^\alpha} W(|\boldsymbol{x} - \boldsymbol{x}_i|, h) d\boldsymbol{x}. \end{aligned} \tag{2.15}$$

Here, we have integrated by parts and used the relation $W(r, h) = W(-r, h)$. From Eq. (2.11), we achieve the following identity:

$$\sum_j \frac{m_j}{\rho(\boldsymbol{x})} W(|\boldsymbol{x} - \boldsymbol{x}_j|, h) = 1. \tag{2.16}$$

By using the identity Eq (2.16), we further modify Eq. (2.15) as

$$\frac{\partial f_i}{\partial x_i^\alpha} = \sum_j \int m_j \frac{f(\boldsymbol{x})}{\rho(\boldsymbol{x})} \frac{\partial}{\partial x_i^\alpha} W(|\boldsymbol{x} - \boldsymbol{x}_i|, h) W(|\boldsymbol{x} - \boldsymbol{x}_j|, h) d\boldsymbol{x}. \tag{2.17}$$

Basically, the spatial derivatives in Eqs. (2.6), (2.8)–(2.10) are evaluated as Eq. (2.17). However, Eq. (2.17) includes the spatial integral and we cannot directly evaluate the gradients using Eq. (2.17).

2.1.3 Equations for Standard SPH Method

In widely-used SPH method, the approximation of $W(|\boldsymbol{x} - \boldsymbol{x}_j|, h) \approx \delta(|\boldsymbol{x} - \boldsymbol{x}_j|)$ is utilized in the evaluation of Eq. (2.17), where $\delta(|\boldsymbol{x} - \boldsymbol{x}_j|)$ is the Dirac δ function. This approximation leads to

$$\frac{\partial f_i}{\partial x_i^\alpha} \approx \sum_j m_j \frac{f_j}{\rho_j} \frac{\partial}{\partial x_i^\alpha} W(|\boldsymbol{x}_i - \boldsymbol{x}_j|, h). \tag{2.18}$$

Then spatial gradients are straightforwardly evaluated using information of neighbor SPH particles as Eq. (2.18).

There are several types of final forms of Eqs. (2.6), (2.8)–(2.10) for the standard SPH method with small differences. Although the final forms depend on detailed procedure how to transform the equations, the final forms of the equations can be basically derived by replacing the spatial derivatives in Eqs. (2.6), (2.8)–(2.10) with those in Eq. (2.18).

For example, Sugiura et al. (2018) use the following types of the equation of continuity, motion, and energy:

$$\frac{d\rho_i}{dt} = -\sum_j m_j \frac{\rho_i}{\rho_j}(v_j^\alpha - v_i^\alpha)\frac{\partial}{\partial x_i^\alpha} W(|\boldsymbol{x}_i - \boldsymbol{x}_j|, h), \tag{2.19}$$

$$\frac{dv_i^\alpha}{dt} = \sum_j m_j \left[\frac{\sigma_i^{\alpha\beta}}{\rho_i^2} + \frac{\sigma_j^{\alpha\beta}}{\rho_j^2} - \Pi_{ij}\delta^{\alpha\beta}\right]\frac{\partial}{\partial x_i^\beta} W(|\boldsymbol{x}_i - \boldsymbol{x}_j|, h), \tag{2.20}$$

$$\begin{aligned} \frac{du_i}{dt} &= -\sum_j \frac{1}{2} m_j \left[\frac{p_i}{\rho_i^2} + \frac{p_j}{\rho_j^2} + \Pi_{ij}\right](v_j^\alpha - v_i^\alpha)\frac{\partial}{\partial x_i^\alpha} W(|\boldsymbol{x}_i - \boldsymbol{x}_j|, h) \\ &+ \sum_j \frac{1}{2} m_j \frac{S_i^{\alpha\beta}}{\rho_i \rho_j}\left[(v_j^\alpha - v_i^\alpha)\frac{\partial}{\partial x_i^\beta} + (v_j^\beta - v_i^\beta)\frac{\partial}{\partial x_i^\alpha}\right] W(|\boldsymbol{x}_i - \boldsymbol{x}_j|, h), \end{aligned} \tag{2.21}$$

where Π_{ij} is the artificial viscosity term, which is necessary to treat shock waves and stabilize numerical simulations. There are many forms of the artificial viscosity term, and the most common form of Π_{ij} is as follows (Monaghan 1992):

$$\begin{aligned} \Pi_{ij} &= \begin{cases} \frac{-\alpha_{\text{vis}}\mu_{ij}(C_{s,i}+C_{s,j})/2+\beta_{\text{vis}}\mu_{ij}^2}{(\rho_i+\rho_j)/2} & (\boldsymbol{v}_i - \boldsymbol{v}_j)\cdot(\boldsymbol{x}_i - \boldsymbol{x}_j) < 0 \\ 0 & (\boldsymbol{v}_i - \boldsymbol{v}_j)\cdot(\boldsymbol{x}_i - \boldsymbol{x}_j) > 0 \end{cases}, \\ \mu_{ij} &= \frac{h(\boldsymbol{v}_i - \boldsymbol{v}_j)\cdot(\boldsymbol{x}_i - \boldsymbol{x}_j)}{(\boldsymbol{x}_i - \boldsymbol{x}_j)^2 + 0.01h^2}. \end{aligned} \tag{2.22}$$

Here, α_{vis} and β_{vis} are the parameters for the artificial viscosity, and throughout this dissertation we use $\alpha_{\text{vis}} = 1.0$ and $\beta_{\text{vis}} = 2.0$.

The density is calculated using both Eqs. (2.11) and (2.19). However, Eq. (2.11) causes problems around surfaces of solid bodies. The density around free surfaces calculated through Eq. (2.11) is smaller than uncompressed density, and the pressure becomes negative through the equation of state. Thus such unphysical pressure around free surfaces deforms solid bodies (Monaghan 1988). The problem is prevented by setting initial densities of all SPH particles to undeformed density and calculating the density using Eq. (2.19), which represents the time-evolution equation for the density.

Sugiura et al. (2018) use the following form of the time-evolution equation of the deviatoric stress tensor:

$$\frac{dS_i^{\alpha\beta}}{dt} = 2\mu\left(\epsilon_i^{\alpha\beta} - \frac{1}{3}\epsilon_i^{\gamma\gamma}\delta^{\alpha\beta}\right) + S_i^{\alpha\gamma}R_i^{\beta\gamma} + S_i^{\beta\gamma}R_i^{\alpha\gamma}, \tag{2.23}$$

where

$$\epsilon_i^{\alpha\beta} = \frac{1}{2}\left(\frac{\partial v_i^\alpha}{\partial x_i^\beta} + \frac{\partial v_i^\beta}{\partial x_i^\alpha}\right), \tag{2.24}$$

$$R_i^{\alpha\beta} = \frac{1}{2}\left(\frac{\partial v_i^\alpha}{\partial x_i^\beta} - \frac{\partial v_i^\beta}{\partial x_i^\alpha}\right), \tag{2.25}$$

and the velocity gradients are evaluated as

$$\frac{\partial v_i^\alpha}{\partial x_i^\beta} = \sum_j \frac{m_j}{\rho_j}(v_j^\alpha - v_i^\alpha)\mathrm{L}_i^{\beta\gamma}\frac{\partial}{\partial x_i^\gamma}W(|\boldsymbol{x}_i - \boldsymbol{x}_j|, h),$$

$$\mathbf{L}_i = \left(\sum_j \frac{m_j}{\rho_j}\frac{\partial}{\partial \boldsymbol{x}_i}W(|\boldsymbol{x}_i - \boldsymbol{x}_j|, h) \otimes (\boldsymbol{x}_j - \boldsymbol{x}_i)\right)^{-1}. \tag{2.26}$$

Here, $\mathbf{L}_i$ is the correction matrix necessary to describe rigid body rotations correctly (Bonet and Lok 1999), and $\otimes$ represents the tensor product. In the rest of this subsection, we explain why we need correction matrix $\mathbf{L}_i$.

The velocity gradient is also calculated without the correction matrix as follows (Libersky and Petschek 1991):

$$\frac{\partial v_i^\alpha}{\partial x_i^\beta} = \sum_j \frac{m_j}{\rho_j}(v_j^\alpha - v_i^\alpha)\frac{\partial}{\partial x_i^\beta}W(|\boldsymbol{x}_i - \boldsymbol{x}_j|, h). \tag{2.27}$$

However, this equation leads to erroneous velocity gradients around surfaces of bodies. To show this, we consider a body doing a rigid body rotation with an angular velocity vector $\boldsymbol{\omega} = (\omega_x, \omega_y, \omega_z)$ and the center of the rotation at the origin of a coordinate. The velocity at the position $\boldsymbol{x}$ is expressed as

$$\boldsymbol{v} = \boldsymbol{\omega} \times \boldsymbol{x}. \tag{2.28}$$

Here, we define the velocity gradient matrix $\mathbf{W}$ so that $\alpha\beta$ component of $\mathbf{W}$ has $\partial v^\alpha/\partial x^\beta$. For the rigid body rotation, $\mathbf{W}$ is expressed as

$$\mathbf{W} = \begin{bmatrix} 0 & -\omega_z & \omega_y \\ \omega_z & 0 & -\omega_x \\ -\omega_y & \omega_x & 0 \end{bmatrix}. \tag{2.29}$$

We also notice the following relationship:

$$\boldsymbol{v} = \mathbf{W}\boldsymbol{x}. \tag{2.30}$$

The velocity gradient evaluated by the SPH approximation Eq. (2.27) becomes

$$\mathbf{W}_{\mathrm{SPH}} = \sum_j \frac{m_j}{\rho_j}(\mathbf{W}\boldsymbol{x}_j - \mathbf{W}\boldsymbol{x}_i) \otimes \frac{\partial}{\partial \boldsymbol{x}_i} W(|\boldsymbol{x}_i - \boldsymbol{x}_j|, h)$$
$$= \mathbf{W}\left(\sum_j \frac{m_j}{\rho_j}(\boldsymbol{x}_j - \boldsymbol{x}_i) \otimes \frac{\partial}{\partial \boldsymbol{x}_i} W(|\boldsymbol{x}_i - \boldsymbol{x}_j|, h)\right), \tag{2.31}$$

where $\partial/\partial \boldsymbol{x}_i = (\partial/\partial x_i, \partial/\partial y_i, \partial/\partial z_i)$. The velocity gradient obtained from the SPH approximation $\mathbf{W}_{\mathrm{SPH}}$ matches the correct velocity gradient $\mathbf{W}$ only when

$$\sum_j \frac{m_j}{\rho_j}(\boldsymbol{x}_j - \boldsymbol{x}_i) \otimes \frac{\partial}{\partial \boldsymbol{x}_i} W(|\boldsymbol{x}_i - \boldsymbol{x}_j|, h) = \mathbf{I}, \tag{2.32}$$

where $\mathbf{I}$ is an identity matrix. However, Eq. (2.32) is not satisfied around surfaces of bodies, which leads to the incorrect treatment of rigid body rotations.

To solve this problem, Bonet and Lok (1999) introduced the correction matrix $\mathbf{L}_i$ and they replaced $(\partial/\partial \boldsymbol{x}_i) W(|\boldsymbol{x}_i - \boldsymbol{x}_j|, h)$ with $\mathbf{L}_i (\partial/\partial \boldsymbol{x}_i) W(|\boldsymbol{x}_i - \boldsymbol{x}_j|, h)$. This replacement modifies Eq. (2.32) as

$$\sum_j \frac{m_j}{\rho_j}(\boldsymbol{x}_j - \boldsymbol{x}_i) \otimes \mathbf{L}_i \frac{\partial}{\partial \boldsymbol{x}_i} W(|\boldsymbol{x}_i - \boldsymbol{x}_j|, h)$$
$$= \left(\sum_j \frac{m_j}{\rho_j}(\boldsymbol{x}_j - \boldsymbol{x}_i) \otimes \frac{\partial}{\partial \boldsymbol{x}_i} W(|\boldsymbol{x}_i - \boldsymbol{x}_j|, h)\right) \mathbf{L}_i^{\mathrm{T}} = \mathbf{I}. \tag{2.33}$$

To satisfy Eq. (2.33), $\mathbf{L}_i$ should be determined as in Eq. (2.26). Introduction of $\mathbf{L}_i$ leads to the equation of the velocity gradient Eq. (2.26) and we can correctly describe rigid body rotations.

To show the validity of the correction matrix, we conduct a simple test simulation of the rotation of an elastic sphere. We model the elastic sphere using about 500 SPH particles. Here, we use the following unit system: The density is scaled using the undeformed density ρ_0, the velocity is scaled using the bulk sound speed C_s, and the length is scaled using the radius of the sphere R_s. The shear modulus is set to $\mu = 1.0 C_s^2 \rho_0$. We initially put a rigid body rotation around z-axis with an angular frequency of $\omega_z = (2\pi/100)(C_s/R_s)$ to the sphere, and calculate the motion of the sphere until one rotation.

Figure 2.1 shows the results of the simulations with and without the correction matrix. The result of the simulation without the correction matrix shows that the rotation speed decreases with time and the sphere finally does a counter rotation. This nonconservation of the rotation speed or the angular momentum is caused by unphysical torque due to the error of the velocity gradient. In contrast, the result of the simulation with the correction matrix shows that the rotation speed of the sphere is well conserved at least until one rotation. Thus the correction matrix is valid for rigid body rotation problems.

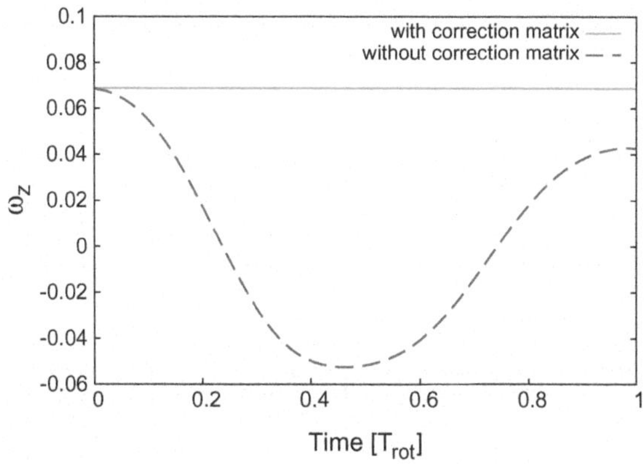

Fig. 2.1 Time evolution of the angular frequency of the rotating elastic sphere calculated using the standard SPH method. The horizontal axis represents the elapsed time normalized by the initial rotation period, and the vertical axis represents the angular frequency of the sphere. The red solid curve shows the simulation with the correction matrix (Eq. 2.26), and the blue dashed curve shows the simulation without the correction matrix (Eq. 2.27)

2.1.4 Equations for Godunov SPH Method

There is another type of SPH method, that is, the Godunov SPH method (Inutsuka 2002; Sugiura and Inutsuka 2017), which utilizes so-called Riemann solver to accurately resolve extremely strong shock waves. Another prominent feature of the Godunov SPH method is that they evaluate the convolution Eq. (2.17) more correctly by interpolating the distribution of the density and other physical quantities, which leads to solve a numerical instability (Sugiura and Inutsuka 2016). Here, we introduce the equations of the Godunov SPH method and the Riemann solver.

2.1.4.1 Formulation of Godunov SPH Method

The equation of continuity is the same as that for the standard SPH method (Eq. 2.19) because the difference of the equation of continuity does not affect the stability of numerical simulations (see Sugiura and Inutsuka 2017).

We use the Gaussian kernel Eq. (2.12) for the Godunov SPH method, and the formulation of the Godunov SPH method is based on the Gaussian kernel.

The equation of motion is defined by the convolution of Eq. (2.8). The acceleration of the ith particle is expressed as

$$\frac{d{v_i}^{\alpha}}{dt} \equiv \int \frac{dv^{\alpha}(\boldsymbol{x})}{dt} W(|\boldsymbol{x} - \boldsymbol{x}_i|, h) d\boldsymbol{x} = \int \frac{1}{\rho(\boldsymbol{x})} \frac{\partial}{\partial x^{\beta}} \sigma^{\alpha\beta}(\boldsymbol{x}) W(|\boldsymbol{x} - \boldsymbol{x}_i|, h) d\boldsymbol{x}. \tag{2.34}$$

The right hand side of Eq. (2.34) is transformed using Eq. (2.16) and integration by parts as

$$\frac{dv_i^\alpha}{dt} = \sum_j m_j \int \frac{\sigma^{\alpha\beta}(\boldsymbol{x})}{\rho^2(\boldsymbol{x})}\left[\frac{\partial}{\partial x_i^\beta} - \frac{\partial}{\partial x_j^\beta}\right] W(|\boldsymbol{x}-\boldsymbol{x}_i|, h) W(|\boldsymbol{x}-\boldsymbol{x}_j|, h) d\boldsymbol{x}. \tag{2.35}$$

Similarly, the equation of energy is defined by the convolution of Eq. (2.10) and is expressed as

$$\frac{du_i}{dt} \equiv \int \frac{du(\boldsymbol{x})}{dt} W(|\boldsymbol{x}-\boldsymbol{x}_i|, h) d\boldsymbol{x} = \int \frac{1}{\rho(\boldsymbol{x})} \sigma^{\alpha\beta}(\boldsymbol{x}) \frac{\partial}{\partial x^\beta} v^\alpha(\boldsymbol{x}) W(|\boldsymbol{x}-\boldsymbol{x}_i|, h) d\boldsymbol{x}. \tag{2.36}$$

The right hand side of Eq. (2.36) is transformed through integration by parts as

$$\begin{aligned}
&\int \frac{1}{\rho(\boldsymbol{x})} \sigma^{\alpha\beta}(\boldsymbol{x}) \frac{\partial}{\partial x^\beta} v^\alpha(\boldsymbol{x}) W(|\boldsymbol{x}-\boldsymbol{x}_i|, h) d\boldsymbol{x} \\
&= \int \frac{1}{\rho(\boldsymbol{x})} \frac{\partial}{\partial x^\beta} (\sigma^{\alpha\beta}(\boldsymbol{x}) v^\alpha(\boldsymbol{x})) W(|\boldsymbol{x}-\boldsymbol{x}_i|, h) d\boldsymbol{x} \\
&+ \int \frac{1}{\rho(\boldsymbol{x})} v^\alpha(\boldsymbol{x}) \frac{\partial}{\partial x^\beta} \sigma^{\alpha\beta}(\boldsymbol{x}) W(|\boldsymbol{x}-\boldsymbol{x}_i|, h) d\boldsymbol{x}.
\end{aligned} \tag{2.37}$$

Here, we use the following approximation:

$$\begin{aligned}
&\int \frac{1}{\rho(\boldsymbol{x})} v^\alpha(\boldsymbol{x}) \frac{\partial}{\partial x^\beta} \sigma^{\alpha\beta}(\boldsymbol{x}) W(|\boldsymbol{x}-\boldsymbol{x}_i|, h) d\boldsymbol{x} \\
&\approx \int \frac{1}{\rho(\boldsymbol{x})} v_i^\alpha \frac{\partial}{\partial x^\beta} \sigma^{\alpha\beta}(\boldsymbol{x}) W(|\boldsymbol{x}-\boldsymbol{x}_i|, h) d\boldsymbol{x}.
\end{aligned} \tag{2.38}$$

Using Eqs. (2.16), (2.38), and integration by parts, we can transform Eq. (2.37) into

$$\begin{aligned}
\frac{du_i}{dt} &= \sum_j m_j \int \frac{\sigma^{\alpha\beta}(\boldsymbol{x})}{\rho^2(\boldsymbol{x})} [v^\alpha(\boldsymbol{x}) - v_i^\alpha] \\
&\left[\frac{\partial}{\partial x_i^\beta} - \frac{\partial}{\partial x_j^\beta}\right] W(|\boldsymbol{x}-\boldsymbol{x}_i|, h) W(|\boldsymbol{x}-\boldsymbol{x}_j|, h) d\boldsymbol{x}.
\end{aligned} \tag{2.39}$$

Finally, we formulate the time-development equation for $S^{\alpha\beta}/\rho$ according to the formulation of the induction equation for magnetic field described in Iwasaki and Inutsuka (2011). We simply differentiate $S^{\alpha\beta}/\rho$ and obtain

$$\frac{d}{dt}\left(\frac{S^{\alpha\beta}}{\rho}\right) = \frac{1}{\rho}\frac{dS^{\alpha\beta}}{dt} - \frac{S^{\alpha\beta}}{\rho^2}\frac{d\rho}{dt}. \tag{2.40}$$

We substitute Eqs. (2.6) and (2.9) into Eq. (2.40) and obtain

$$\frac{d}{dt}\left(\frac{S^{\alpha\beta}}{\rho}\right) = 2\mu\left(\frac{\dot{\epsilon}^{\alpha\beta}}{\rho} - \frac{1}{3}\delta^{\alpha\beta}\frac{\dot{\epsilon}^{\gamma\gamma}}{\rho}\right) + \frac{S^{\alpha\gamma}}{\rho}R^{\beta\gamma} + \frac{S^{\beta\gamma}}{\rho}R^{\alpha\gamma} + \frac{S^{\alpha\beta}}{\rho}\dot{\epsilon}^{\gamma\gamma}. \quad (2.41)$$

Note that $(\partial/\partial x^{\gamma})v^{\gamma} = \dot{\epsilon}^{\gamma\gamma}$. We again define the time derivative of $S^{\alpha\beta}/\rho$ of the ith particle as the convolution of Eq. (2.41):

$$\begin{aligned}\frac{d}{dt}\left(\frac{S^{\alpha\beta}}{\rho}\right)_i &\equiv \int \left[\frac{d}{dt}\left(\frac{S^{\alpha\beta}}{\rho}\right)\right](\boldsymbol{x})W(|\boldsymbol{x}-\boldsymbol{x}_i|, h)d\boldsymbol{x} \\ &= \int \left[2\mu\left(\frac{\dot{\epsilon}^{\alpha\beta}(\boldsymbol{x})}{\rho(\boldsymbol{x})} - \frac{1}{3}\delta^{\alpha\beta}\frac{\dot{\epsilon}^{\gamma\gamma}(\boldsymbol{x})}{\rho(\boldsymbol{x})}\right) + \frac{S^{\alpha\gamma}(\boldsymbol{x})}{\rho(\boldsymbol{x})}R^{\beta\gamma}(\boldsymbol{x})\right. \\ &\quad \left. + \frac{S^{\beta\gamma}(\boldsymbol{x})}{\rho(\boldsymbol{x})}R^{\alpha\gamma}(\boldsymbol{x}) + \frac{S^{\alpha\beta}(\boldsymbol{x})}{\rho(\boldsymbol{x})}\dot{\epsilon}^{\gamma\gamma}(\boldsymbol{x})\right]W(|\boldsymbol{x}-\boldsymbol{x}_i|, h)d\boldsymbol{x}. \quad (2.42)\end{aligned}$$

The right hand side of Eq. (2.42) is composed of the following terms ($\dot{\epsilon}^{\alpha\beta}$ and $R^{\alpha\beta}$ are represented by the sums of velocity gradients):

$$\int \frac{1}{\rho(\boldsymbol{x})}\frac{\partial v^{\alpha}(\boldsymbol{x})}{\partial x^{\beta}}W(|\boldsymbol{x}-\boldsymbol{x}_i|, h)d\boldsymbol{x}, \quad (2.43)$$

$$\int \frac{S^{\alpha'\beta'}(\boldsymbol{x})}{\rho(\boldsymbol{x})}\frac{\partial v^{\alpha}(\boldsymbol{x})}{\partial x^{\beta}}W(|\boldsymbol{x}-\boldsymbol{x}_i|, h)d\boldsymbol{x}, \quad (2.44)$$

where α, β, α', and β' show arbitrary direction appearing in each term of Eq. (2.42). We notice that Eq. (2.44) is almost the same as the right hand side of Eq. (2.36), and thus we can do the same transformation as Eqs. (2.37)–(2.39):

$$\begin{aligned}&\int \frac{S^{\alpha'\beta'}(\boldsymbol{x})}{\rho(\boldsymbol{x})}\frac{\partial v^{\alpha}(\boldsymbol{x})}{\partial x^{\beta}}W(|\boldsymbol{x}-\boldsymbol{x}_i|, h)d\boldsymbol{x} \\ &= \sum_j m_j \int \frac{S^{\alpha'\beta'}(\boldsymbol{x})}{\rho^2(\boldsymbol{x})}[v^{\alpha}(\boldsymbol{x}) - v_i^{\alpha}]\left[\frac{\partial}{\partial x_i^{\beta}} - \frac{\partial}{\partial x_j^{\beta}}\right]W(|\boldsymbol{x}-\boldsymbol{x}_i|, h)W(|\boldsymbol{x}-\boldsymbol{x}_j|, h)d\boldsymbol{x}.\end{aligned} \quad (2.45)$$

Using Eqs. (2.11), (2.16), $\partial v_i^{\alpha}/\partial x^{\beta} = 0$, and the partial integration, Eq. (2.43) is transformed as

$$\begin{aligned}&\int \frac{1}{\rho(\boldsymbol{x})}\frac{\partial v^{\alpha}(\boldsymbol{x})}{\partial x^{\beta}}W(|\boldsymbol{x}-\boldsymbol{x}_i|, h)d\boldsymbol{x} \\ &= \int \frac{1}{\rho(\boldsymbol{x})}\frac{\partial}{\partial x^{\beta}}(v^{\alpha}(\boldsymbol{x}) - v_i^{\alpha})W(|\boldsymbol{x}-\boldsymbol{x}_i|, h)d\boldsymbol{x}\end{aligned}$$

$$
\begin{aligned}
&= -\int (v^\alpha(\boldsymbol{x}) - v_i^\alpha)\frac{\partial}{\partial x^\beta}\left(\frac{W(|\boldsymbol{x}-\boldsymbol{x}_i|, h)}{\rho(\boldsymbol{x})}\right)d\boldsymbol{x} \\
&= \sum_j m_j \int \frac{1}{\rho^2(\boldsymbol{x})}(v^\alpha(\boldsymbol{x}) - v_i^\alpha)\left[\frac{\partial}{\partial x_i^\beta} - \frac{\partial}{\partial x_j^\beta}\right]W(|\boldsymbol{x}-\boldsymbol{x}_i|, h)W(|\boldsymbol{x}-\boldsymbol{x}_j|, h)d\boldsymbol{x}.
\end{aligned}
\tag{2.46}
$$

To evaluate the integrals in Eqs. (2.35), (2.39), (2.45), and (2.46), we make interpolated functions for physical quantities, especially the density. Here, a new convenient coordinate system is defined for the integration. We have the origin of the coordinate system at $(\boldsymbol{x}_i + \boldsymbol{x}_j)/2$, and the s-axis is defined to be parallel to the vector $\boldsymbol{x}_i - \boldsymbol{x}_j$. $\boldsymbol{s}_\perp$ denotes the component perpendicular to the s-axis of a vector $\boldsymbol{x}$. The vector $\boldsymbol{e}_{ij} \equiv (\boldsymbol{x}_i - \boldsymbol{x}_j)/|\boldsymbol{x}_i - \boldsymbol{x}_j|$ is defined to be the unit vector parallel to the s-axis, and $\Delta s_{ij} \equiv |\boldsymbol{x}_i - \boldsymbol{x}_j|$ as the distance between the ith and jth particles.

Equations (2.35), (2.39), (2.45), and (2.46) are essentially composed of the integration shown as

$$
\int \frac{f(\boldsymbol{x})}{\rho^2(\boldsymbol{x})}W(|\boldsymbol{x}-\boldsymbol{x}_i|, h)W(|\boldsymbol{x}-\boldsymbol{x}_j|, h)d\boldsymbol{x}, \tag{2.47}
$$

where $f(\boldsymbol{x})$ is an arbitrary physical quantity. Here, the weighted average of the physical quantity f_{ij}^* is defined as

$$
\begin{aligned}
&\int \frac{f(\boldsymbol{x})}{\rho^2(\boldsymbol{x})}W(|\boldsymbol{x}-\boldsymbol{x}_i|, h)W(|\boldsymbol{x}-\boldsymbol{x}_j|, h)d\boldsymbol{x} \\
&= f_{ij}^* \int \frac{1}{\rho^2(\boldsymbol{x})}W(|\boldsymbol{x}-\boldsymbol{x}_i|, h)W(|\boldsymbol{x}-\boldsymbol{x}_j|, h)d\boldsymbol{x}.
\end{aligned}
\tag{2.48}
$$

If we interpolate $f(\boldsymbol{x})$ linearly in the direction parallel to $\boldsymbol{e}_{ij}$ and constant in the direction perpendicular to $\boldsymbol{e}_{ij}$, f_{ij}^* is expressed as

$$
f_{ij}^* = \frac{f_i + f_j}{2} + s_{ij}^* \frac{f_i - f_j}{|\boldsymbol{x}_i - \boldsymbol{x}_j|}, \tag{2.49}
$$

$$
s_{ij}^* = \int_{-\infty}^{\infty} \frac{\sqrt{2}}{h\sqrt{\pi}} \frac{s}{\rho^2(s)} \exp\left(-\frac{2s^2}{h^2}\right)ds. \tag{2.50}
$$

Then, $\rho^{-2}(\boldsymbol{r})$ is expanded linearly in the direction perpendicular to $\boldsymbol{e}_{ij}$ as

$$
\rho^{-2}(\boldsymbol{x}) \approx \rho^{-2}(s) + \boldsymbol{s}_\perp \cdot \nabla\rho^{-2}(s). \tag{2.51}
$$

Note that the second term of Eq. (2.51) vanishes through the integration due to the symmetric property of the kernel function. Equation (2.48) is transformed using Eq. (2.51) as

$$\int \frac{f(\boldsymbol{x})}{\rho^2(\boldsymbol{x})} W(|\boldsymbol{x}-\boldsymbol{x}_i|, h) W(|\boldsymbol{x}-\boldsymbol{x}_j|, h) d\boldsymbol{x} = f^*_{ij} V^2_{ij} W(|\boldsymbol{x}_i - \boldsymbol{x}_j|, \sqrt{2}h), \quad (2.52)$$

where

$$V^2_{ij} = \int_{-\infty}^{\infty} \frac{\sqrt{2}}{h\sqrt{\pi}} \frac{1}{\rho^2(s)} \exp\left(-\frac{2s^2}{h^2}\right) ds. \quad (2.53)$$

To evaluate the integrals of Eqs. (2.50) and (2.53), we interpolate $1/\rho(s)$ using the densities and its gradients at the ith and jth particles. For the convenience, we define the specific volume as $V(s) = 1/\rho(s)$ and calculate the specific volume and its gradients at the ith particle as

$$\begin{aligned} V_i &= \frac{1}{\rho_i}, \\ \frac{\partial V_i}{\partial x_i^\alpha} &= -\frac{1}{\rho_i^2} \frac{\partial \rho_i}{\partial x_i^\alpha} = -\frac{1}{\rho_i^2} \sum_j m_j \frac{\partial}{\partial x_i^\alpha} W(|\boldsymbol{x}_i - \boldsymbol{x}_j|, h), \\ \frac{\partial^2 V_i}{\partial x_i^\alpha \partial x_i^\beta} &= \sum_j \frac{m_j}{\rho_j} \frac{\partial V_j}{\partial x_j^\alpha} \frac{\partial}{\partial x_i^\beta} W(|\boldsymbol{x}_i - \boldsymbol{x}_j|, h). \end{aligned} \quad (2.54)$$

We prepare three interpolation methods, that is, the linear, the cubic spline, and the quintic spline interpolation. For the linear interpolation, $V(s)$ is expressed as

$$V(s) = C_{ij} s + D_{ij}, \quad (2.55)$$

where

$$\begin{aligned} C_{ij} &= \frac{V_i - V_j}{\Delta s_{ij}}, \\ D_{ij} &= \frac{V_i + V_j}{2}. \end{aligned} \quad (2.56)$$

Then we achieve the expressions of V^2_{ij} and s^*_{ij} for the linear interpolation as

$$\begin{aligned} V^2_{ij,\text{linear}} &= \frac{1}{4} h^2 C^2_{ij} + D^2_{ij}, \\ s^*_{ij,\text{linear}} &= \frac{h^2 C_{ij} D_{ij}}{2 V^2_{ij,\text{linear}}}. \end{aligned} \quad (2.57)$$

For the cubic spline interpolation, $V(s)$ is expressed as

$$V(s) = A_{ij} s^3 + B_{ij} s^2 + C_{ij} s + D_{ij}, \quad (2.58)$$

where

$$
\begin{aligned}
A_{ij} &= -2\frac{V_i - V_j}{\Delta s_{ij}^3} + \frac{V_i^{'} + V_j^{'}}{\Delta s_{ij}^2}, \\
B_{ij} &= \frac{1}{2}\frac{V_i^{'} - V_j^{'}}{\Delta s_{ij}}, \\
C_{ij} &= \frac{3}{2}\frac{V_i - V_j}{\Delta s_{ij}} - \frac{1}{4}(V_i^{'} + V_j^{'}), \\
D_{ij} &= \frac{1}{2}(V_i + V_j) - \frac{1}{8}(V_i^{'} - V_j^{'})\Delta s_{ij}, \\
V_i^{'} &= e_{ij}^{\alpha}\frac{\partial V_i}{\partial x_i^{\alpha}}, \\
V_j^{'} &= e_{ij}^{\alpha}\frac{\partial V_j}{\partial x_j^{\alpha}}.
\end{aligned} \tag{2.59}
$$

Then we achieve the expressions of V_{ij}^2 and s_{ij}^* for the cubic spline interpolation as

$$
\begin{aligned}
V_{ij,\mathrm{cubic}}^2 &= \frac{15}{64}h^6 A_{ij}^2 + \frac{3}{16}h^4(2A_{ij}C_{ij} + B_{ij}^2) + \frac{1}{4}h^2(2B_{ij}D_{ij} + C_{ij}^2) + D_{ij}^2, \\
s_{ij,\mathrm{cubic}}^* &= \frac{(15/32)h^6 A_{ij}B_{ij} + (3/8)h^4(A_{ij}D_{ij} + B_{ij}C_{ij}) + (1/2)h^2 C_{ij}D_{ij}}{V_{ij,\mathrm{cubic}}^2}.
\end{aligned} \tag{2.60}
$$

Finally, for the quintic spline interpolation, $V(s)$ is expressed as

$$
V(s) = A_{ij}s^5 + B_{ij}s^4 + C_{ij}s^3 + D_{ij}s^2 + E_{ij}s + F_{ij}, \tag{2.61}
$$

where

$$
\begin{aligned}
A_{ij} &= 6\frac{V_i - V_j}{\Delta s_{ij}^5} - 3\frac{V_i^{'} + V_j^{'}}{\Delta s_{ij}^4} + \frac{1}{2}\frac{V_i^{''} - V_j^{''}}{\Delta s_{ij}^3}, \\
B_{ij} &= -\frac{1}{2}\frac{V_i^{'} - V_j^{'}}{\Delta s_{ij}^3} + \frac{1}{4}\frac{V_i^{''} + V_j^{''}}{\Delta s_{ij}^2}, \\
C_{ij} &= -5\frac{V_i - V_j}{\Delta s_{ij}^3} + \frac{5}{2}\frac{V_i^{'} + V_j^{'}}{\Delta s_{ij}^2} - \frac{1}{4}\frac{V_i^{''} - V_j^{''}}{\Delta s_{ij}}, \\
D_{ij} &= \frac{3}{4}\frac{V_i^{'} - V_j^{'}}{\Delta s_{ij}} - \frac{1}{8}(V_i^{''} + V_j^{''}), \\
E_{ij} &= \frac{15}{8}\frac{V_i - V_j}{\Delta s_{ij}} - \frac{7}{16}(V_i^{'} + V_j^{'}) + \frac{1}{32}(V_i^{''} - V_j^{''})\Delta s_{ij},
\end{aligned}
$$

$$F_{ij} = \frac{1}{2}(V_i + V_j) - \frac{5}{32}(V_i' - V_j')\Delta s_{ij} + \frac{1}{64}(V_i'' + V_j'')\Delta s_{ij}^2, \tag{2.62}$$

and

$$\begin{aligned} V_i'' &= e_{ij}^{\alpha} e_{ij}^{\beta} \frac{\partial^2 V_i}{\partial x_i^{\alpha} \partial x_i^{\beta}}, \\ V_j'' &= e_{ij}^{\alpha} e_{ij}^{\beta} \frac{\partial^2 V_j}{\partial x_j^{\alpha} \partial x_j^{\beta}}. \end{aligned} \tag{2.63}$$

Then we achieve the expressions of V_{ij}^2 and s_{ij}^* for the quintic spline interpolation as

$$\begin{aligned} V_{ij,\text{quintic}}^2 &= \frac{945}{1024} h^{10} A_{ij}^2 + \frac{105}{256} h^8 (2A_{ij} C_{ij} + B_{ij}^2) + \frac{15}{64} h^6 (2A_{ij} E_{ij} + 2B_{ij} D_{ij} + C_{ij}^2) \\ &+ \frac{3}{16} h^4 (2B_{ij} F_{ij} + 2C_{ij} E_{ij} + D_{ij}^2) + \frac{1}{4} h^2 (2D_{ij} F_{ij} + E_{ij}^2) + F_{ij}^2, \\ s_{ij,\text{quintic}}^* &= \left(\frac{945}{512} h^{10} A_{ij} B_{ij} + \frac{105}{128} h^8 (A_{ij} D_{ij} + B_{ij} C_{ij}) + \frac{15}{32} h^6 (A_{ij} F_{ij} + B_{ij} E_{ij} + C_{ij} D_{ij}) \right. \\ &\left. + \frac{3}{8} h^4 (C_{ij} F_{ij} + D_{ij} E_{ij}) + \frac{1}{2} h^2 E_{ij} F_{ij} \right) / V_{ij,\text{quintic}}^2. \end{aligned} \tag{2.64}$$

For the stability of numerical simulations, we should select appropriate interpolation methods from the linear, the cubic spline, and the quintic spline interpolation depending on spatial dimension and the sign of pressure (see Sect. 2.1.5).

Here, we introduce the final forms of the equation of motion, the equation of energy, and the time-evolution equation for the deviatoric stress tensor for the Godunov SPH method. The equation of motion is expressed as

$$\frac{dv_i^{\alpha}}{dt} = \sum_j 2m_j \sigma_{ij}^{\alpha\beta *} V_{ij}^2 \frac{\partial}{\partial x_i^{\beta}} W(|\boldsymbol{x}_i - \boldsymbol{x}_j|, \sqrt{2}h), \tag{2.65}$$

where

$$\begin{aligned} \sigma_{ij}^{\alpha\beta *} &= -P_{ij}^* \delta^{\alpha\beta} + S_{ij}^{\alpha\beta *}, \\ S_{ij}^{\alpha\beta *} &= \frac{S_i^{\alpha\beta} + S_j^{\alpha\beta}}{2} + s_{ij}^* \frac{S_i^{\alpha\beta} - S_j^{\alpha\beta}}{|\boldsymbol{x}_i - \boldsymbol{x}_j|}. \end{aligned} \tag{2.66}$$

The equation of energy is expressed as

$$\frac{du_i}{dt} = \sum_j 2m_j \sigma_{ij}^{\alpha\beta *} V_{ij}^2 (v_{ij}^{\alpha *} - v_i^{\alpha *}) \frac{\partial}{\partial x_i^{\beta}} W(|\boldsymbol{x}_i - \boldsymbol{x}_j|, \sqrt{2}h), \tag{2.67}$$

where

$$v_i^{\alpha *} = v_i^{\alpha} + \frac{1}{2}\Delta t \frac{dv_i^{\alpha}}{dt}, \tag{2.68}$$

for the conservation of the total energy. The time-evolution equation of the deviatoric stress tensor is expressed as

$$\frac{d}{dt}\left(\frac{S^{\alpha\beta}}{\rho}\right)_i = \sum_j 2\mu\left(\dot{\epsilon}^{\alpha\beta}_{\rho,ij} - \frac{1}{3}\delta^{\alpha\beta}\dot{\epsilon}^{\gamma\gamma}_{\rho,ij}\right) + S^{\alpha\gamma *}_{ij} R^{\beta\gamma}_{\rho,ij} + S^{\beta\gamma *}_{ij} R^{\alpha\gamma}_{\rho,ij} + S^{\alpha\beta *}_{ij}\dot{\epsilon}^{\gamma\gamma}_{\rho,ij}, \tag{2.69}$$

where,

$$\dot{\epsilon}^{\alpha\beta}_{\rho,ij} \equiv m_j V_{ij}^2 \Big[(v_{ij}^{\alpha *} - v_i^{\alpha})\frac{\partial}{\partial x_i^{\beta}} + (v_{ij}^{\beta *} - v_i^{\beta})\frac{\partial}{\partial x_i^{\alpha}}\Big] W(|\boldsymbol{x}_i - \boldsymbol{x}_j|, \sqrt{2}h),$$

$$R^{\alpha\beta}_{\rho,ij} \equiv m_j V_{ij}^2 \Big[(v_{ij}^{\alpha *} - v_i^{\alpha})\frac{\partial}{\partial x_i^{\beta}} - (v_{ij}^{\beta *} - v_i^{\beta})\frac{\partial}{\partial x_i^{\alpha}}\Big] W(|\boldsymbol{x}_i - \boldsymbol{x}_j|, \sqrt{2}h). \tag{2.70}$$

Note that the deviatoric stress tensor itself is obtained as $S_i^{\alpha\beta} = (S^{\alpha\beta}/\rho)_i \rho_i$.

We introduced the correction matrix $\mathbf{L}_i$ for the standard SPH method. To correctly treat rigid body rotations for simulations using the Godunov SPH method, we also need such correction matrix for the evaluation of the velocity gradients appearing in Eq. (2.70). However, we have not yet developed the correction matrix for the Godunov SPH method. We will do this in our future works.

According to Inutsuka (2002), we use the pressure resulted from a Riemann problem utilizing the physical values associated with the ith and jth particles for P_{ij}^*. Inutsuka (2002) also utilizes the velocity resulted from the Riemann problem for $v_{ij}^{\alpha *}$. The simulations for hydrodynamics work well with the velocity resulted from the Riemann problem. However, this may cause problems for elastic dynamics because of the difference of equation of states, and it is safe to utilize the simple averaged value for $v_{ij}^{\alpha *}$ that is expressed as

$$v_{ij}^{\alpha *} = \frac{v_i^{\alpha} + v_j^{\alpha}}{2} + s_{ij}^* \frac{v_i^{\alpha} - v_j^{\alpha}}{|\boldsymbol{x}_i - \boldsymbol{x}_j|}. \tag{2.71}$$

2.1.4.2 Riemann Solver

The Riemann problem is a mathematical problem to solve a shock tube problem. The shock tube problem is a physical and experimental problem and the details of the problem are as follows: We firstly prepare a tube with a partition that separates space inside of the tube into left- and right-hand sides. Then we put two different gases with different thermal states to the left- and right-hand sides of the tube. Finally we remove the partition, then a shock wave propagates to lower pressure side, a rarefaction wave propagates to higher pressure side, and an initial contact plane between two different gases propagates as a contact discontinuity. The Riemann problem is the mathematical procedure to solve the shock tube problem. Especially, the Riemann solver is a specific procedure to obtain physical states around the contact discontinuity resulted from the shock tube problem.

We introduce how to calculate P_{ij}^* using the physical quantities associated with the ith and jth particles through the Riemann problem. We only introduce the Riemann solver for the equation of state $P = C_s^2(\rho - \rho_0)$. Please refer to, for example, van Leer (1979) for the Riemann solver for the ideal gas equation of state.

We consider a shock tube problem where the initial state (e.g., pressure or density) of the right hand side is determined by the physical quantities of the ith particle and the initial state of the left hand side is determined by the physical quantities of the jth particle. The quantities with the subscript of R represent the physical quantities at the right hand side, those with the subscript of L represent those at the left hand side, and those with the superscript of $*$ represent those around the contact discontinuity. Thus $P_\mathrm{R} = P_i$, $\rho_\mathrm{R} = \rho_i$, $v_\mathrm{R} = \boldsymbol{v}_i \cdot \boldsymbol{e}_{ij}$, $P_\mathrm{L} = P_j$, $\rho_\mathrm{L} = \rho_j$, and $v_\mathrm{L} = \boldsymbol{v}_j \cdot \boldsymbol{e}_{ij}$. The waves are propagating to both the right and left directions. The waves are the shock wave or the rarefaction wave.

The physical quantities of the pre- and post-shock wave are connected by the following Rankine-Hugoniot relations:

$$\pm W_{s,D}(V^* - V_D) + (v^* - v_D) = 0, \tag{2.72}$$

$$\pm W_{s,D}(v^* - v_D) - (P^* - P_D) = 0, \tag{2.73}$$

where $W_{s,D}$ is the shock speed in the mass coordinate, $V_D = 1/\rho_D$ is the specific volume, the sign represents the direction of the shock propagation, and D also shows the direction of the shock propagation and is replaced by R or L. Equations (2.72), (2.73), and the equation of state for the post-shock medium

$$P^* = C_s^2(\rho^* - \rho_0), \tag{2.74}$$

lead to the equation for the shock speed

$$W_{s,D} = C_s \sqrt{\rho_D \left(\rho_D + \frac{P^* - P_D}{C_s^2} \right)}. \tag{2.75}$$

Across the rarefaction wave, the Riemann invariants $J_\pm = v \pm C_s \ln \rho$ are conserved on the trajectories $dx/dt = v \pm C_s$, and the physical quantities of the pre- and post-rarefaction wave are connected as

$$v_D \mp C_s \ln \rho_D = v^* \mp \ln \rho^*. \tag{2.76}$$

According to the speed of the shock wave Eq. (2.73), we also define $W_{s,D}$ for the rarefaction wave as

$$W_{s,D} \equiv \left|\frac{P^* - P_D}{v^* - v_D}\right| = \frac{P_D - P^*}{C_s}\left[\ln\left(\frac{C_s^2 \rho_D}{P^* + C_s^2 \rho_0}\right)\right]^{-1}, \tag{2.77}$$

where we use Eqs. (2.74) and (2.76).

We rewrite Eqs. (2.73) and (2.77) as follows:

$$-W_{s,\mathrm{L}}(v^* - v_\mathrm{L}) - (P^* - P_\mathrm{L}) = 0, \tag{2.78}$$

$$W_{s,\mathrm{R}}(v^* - v_\mathrm{R}) - (P^* - P_\mathrm{R}) = 0. \tag{2.79}$$

If the shock propagates to D direction, $P^* > P_D$. If the rarefaction wave propagates to D direction, $P^* < P_D$. Thus the speeds of the waves $W_{s,D}$ are represented as

$$W_{s,D} = C_s\sqrt{\rho_D\left(\rho_D + \frac{P^* - P_D}{C_s^2}\right)} \quad (P^* > P_D), \tag{2.80}$$

$$W_{s,D} = \frac{P_D - P^*}{C_s}\left[\ln\left(\frac{C_s^2 \rho_D}{P^* + C_s^2 \rho_0}\right)\right]^{-1} \quad (P^* < P_D). \tag{2.81}$$

From Eqs. (2.78) and (2.79), the pressure P^* and the velocity v^* around the contact discontinuity are written as

$$P^* = \frac{P_\mathrm{L}/W_{s,\mathrm{L}} + P_\mathrm{R}/W_{s,\mathrm{R}} + v_\mathrm{L} - v_\mathrm{R}}{1/W_{s,\mathrm{L}} + 1/W_{s,\mathrm{R}}}, \tag{2.82}$$

$$v^* = \frac{v_\mathrm{L} W_{s,\mathrm{L}} + v_\mathrm{R} W_{s,\mathrm{R}} + P_\mathrm{L} - P_\mathrm{R}}{W_{s,\mathrm{L}} + W_{s,\mathrm{R}}}. \tag{2.83}$$

P^* and v^* are determined through iterative calculations of Eqs. (2.80)–(2.83). Five cycles of the iterations using $P^* = (P_L + P_R)/2$ as the initial value are sufficient for the convergence. Finally, the converged P^* is used for P_{ij}^* in Eqs. (2.65) and (2.67).

However, P^* obtained from the above procedures has the first-order accuracy in space. For the Riemann solver with the second-order accuracy in space, we need to modify P_R, ρ_R, v_R, P_L, ρ_L, and v_L using the gradients of these physical quantities at the ith and jth particles. We skip the explanation of detailed procedures for this modification, and please refer to Inutsuka (2002). We just give a note that the gradient of the quantity f utilized for the Riemann solver should be calculated with

$$\frac{\partial f_i}{\partial x_i^\alpha} = \sum_j \frac{m_j(f_j - f_i)}{\rho_j}\frac{\partial}{\partial x_i^\alpha} W(|\boldsymbol{x}_i - \boldsymbol{x}_j|, h) \tag{2.84}$$

to avoid unphysical gradients nearby free surfaces.

We explained the Riemann solver for the equation of state for elastic bodies $P = C_s^2(\rho - \rho_0)$. However, impact simulations with rocky material utilize other equation of states such as the Tillotson equation of state. In spite of the difference of equation of states, usual equation of states for rocky bodies behave like the equation of state for elastic bodies if the compression and the temperature raise due to impacts are moderate. Thus we can sill utilize the Riemann solver for $P = C_s^2(\rho - \rho_0)$ even if we utilize other equation of states for rocky material. In this case, we approximate C_s and ρ_0 utilized for the Riemann solver using the physical values associated with the ith and jth particles. A possible way to calculate C_s and ρ_0 is as follows:

$$C_s = \frac{C_{s,i} + C_{s,j}}{2},$$
$$\rho_0 = \frac{1}{2}[(\rho_i - P_i/C_s^2) + (\rho_j - P_j/C_s^2)]. \tag{2.85}$$

2.1.5 *Godunov SPH Method and Tensile Instability*

The SPH method has one serious problem especially in simulations of pure elastic dynamics; the tensile instability (e.g., Swegle et al. 1995). The tensile instability is a numerical instability that is prominent in tension-dominated regions, or negative pressure regions and produces unphysical clustering of SPH particles, which leads to unphysical fracturing of elastic bodies (see Gray et al. 2001; Swegle et al. 1995). Thus, we need to suppress the tensile instability for the correct treatment of pure elastic bodies.

Several methods have been proposed to solve the tensile instability (e.g., Chen et al. 1999; Randles and Libersky 1996; Sigalotti and López 2008). The most common way to remove the tensile instability is the method utilizing an artificial stress (Gray et al. 2001; Monaghan 2000). The artificial stress is an additional repulsive force that stabilizes perturbations with short wavelengths comparable to particle spacing. Perturbations with longer wavelengths are not affected by the artificial stress, and thus it safely removes the tensile instability, which is the instability with short wavelength of $2h$.

We develop another way to remove the tensile instability based on the Godunov SPH method. We investigate the stability of Eq. (2.65) through the linear stability analysis (Sugiura and Inutsuka 2016). As a result, we find that the stability depends on spatial dimensions, the sign of pressure, and the interpolation methods for V_{ij}^2 and s_{ij}^*. Thus we can avoid the tensile instability if we use appropriate interpolation methods depending on the sign of pressure. Here, we briefly introduce the results of the linear stability analysis for the Godunov SPH method. See appendix of this chapter for the details of the linear stability analysis.

2.1.5.1 Results of Linear Stability Analysis: One-Dimensional Case

Firstly, we introduce the results of the linear stability analysis for a one-dimensional case. One dimension means that we only put SPH particles on one-dimensional string and use the Gaussian kernel for one dimension, i.e., we set $d_{\mathrm{dim}} = 1$ in Eq. (2.12).

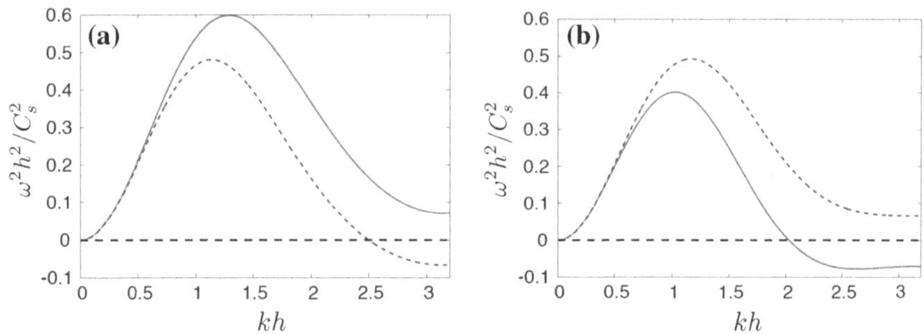

Fig. 2.2 Dispersion relations (DRs) for the Godunov SPH method in a one-dimensional geometry. In both panels, the horizontal axis represents the wave number k of perturbations normalized by the smoothing length h, the vertical axis represents the square of the angular frequency ω of perturbations normalized by the sound speed C_s and the smoothing length h, the red solid curve shows DR for the linear interpolation, the green dotted curve shows that for the cubic spline interpolation, and the blue dashed curve shows that for the quintic spline interpolation. The panel **a** shows the DRs with the positive unperturbed pressure of $\overline{P} = 0.2C_s^2\overline{\rho}$, and the panel **b** shows those with the negative unperturbed pressure of $\overline{P} = -0.2C_s^2\overline{\rho}$. Reprinted from Sugiura and Inutsuka (2016), Copyright (2015), with permission from Elsevier

Figure 2.2 shows the dispersion relations (DRs) for three interpolation methods with positive unperturbed pressure (Fig. 2.2a) and negative unperturbed pressure (Fig. 2.2b). As we notice from Fig. 2.2a, ω^2 for the linear and cubic spline interpolations with the positive pressure is always positive for all wave numbers k including the Nyquist frequency $hk = \pi$, while ω^2 for the quintic spline interpolation with the positive pressure becomes negative around the Nyquist frequency. Note that the perturbations with negative ω^2 are unstable, since ω becomes pure imaginary and then the perturbations exponentially grow (see Eq. 2.126). Therefore, the linear and cubic spline interpolations are stable with positive pressure, while the quintic spline interpolation is unstable with positive pressure. In contrast, Fig. 2.2 shows that the linear and cubic spline interpolation are unstable with negative pressure, while the quintic spline interpolation is stable with negative pressure.

2.1.5.2 Results of Linear Stability Analysis: Two- and Three-Dimensional Case

Secondly, we introduce the results of the linear stability analysis for two- and three-dimensional cases. Again, two dimensions mean that we put SPH particles on two-dimensional plane and use the Gaussian kernel for two dimensions, i.e., we set $d_{\rm dim} = 2$ in Eq. (2.12). Three dimensions mean usual setup of simulations.

The terms a and b (Eq. 2.136) in the DR for the linear interpolation have the same values independent of the number of the spatial dimension. Thus the stability for the linear interpolation is also independent of the number of the spatial dimension;

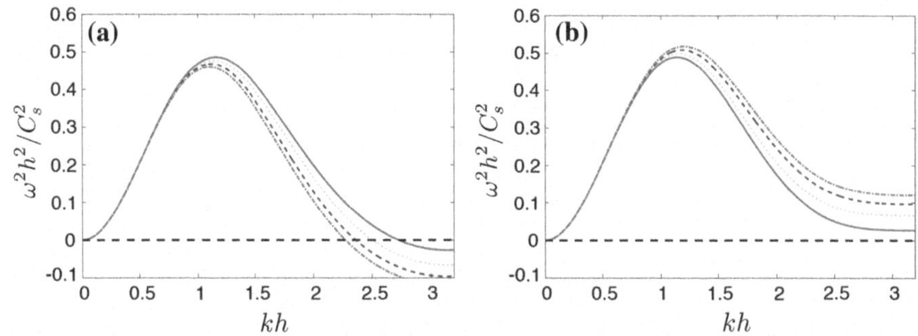

Fig. 2.3 DRs for the Godunov SPH method in two- and three-dimensional geometries. In both panels, the horizontal axis represents the wave number k of perturbations normalized by the smoothing length h, the vertical axis represents the square of the angular frequency ω of perturbations normalized by the sound speed C_s and the smoothing length h, the red solid curve shows the DR for the cubic spline interpolation in the two-dimensional case, the green dotted curve shows that in the three-dimensional case, the blue dashed curve shows the DR for the quintic spline interpolation in two-dimensional case, and the magenta chain curve shows that in the three-dimensional case. The panel **a** shows the DRs with the positive unperturbed pressure of $\overline{P} = 0.2C_s^2\overline{\rho}$, and the panel **b** shows those with the negative unperturbed pressure of $\overline{P} = -0.2C_s^2\overline{\rho}$. Reprinted from Sugiura and Inutsuka (2016), Copyright (2015), with permission from Elsevier

even for the two- and three-dimensional cases, the linear interpolation with positive pressure is stable while that with negative pressure is unstable.

Figure 2.3 shows the DRs for the cubic and quintic spline interpolations in the two- and three-dimensional geometries with the positive pressure (Fig. 2.3a) and the negative pressure (Fig. 2.3b). As we notice from Fig. 2.3a, both cubic and quintic spline interpolations in both two- and three-dimensional geometries are unstable with the positive pressure. In contrast, Fig. 2.3b shows that both cubic and quintic spline interpolations in both two- and three-dimensional geometries are stable with the negative pressure. Therefore, in the two- and three-dimensional cases, the cubic and quintic spline interpolations are unstable with positive pressure and stable with negative pressure.

2.1.5.3 Suppression of Tensile Instability with Godunov SPH Method

We summarize the stability of the Godunov SPH method in Table 2.1. This result suggests that if we select appropriate interpolation methods depending on the sign of pressure, we can avoid the tensile instability. Here, we propose the following way to select the interpolation methods for V_{ij}^2 in the equation of motion (2.65), the equation of energy (2.67), and the time-evolution equation for $S^{\alpha\beta}$ (2.69): V_{ij}^2 in one dimension is

Table 2.1 Dependence of the stability on the spatial dimension, the interpolation methods, and the sing of pressure. Circles (◯) show stable interpolations and crosses (×) show unstable interpolations. $d_{\rm dim}$ shows the number of the dimensions and P shows the pressure. Reprinted from Sugiura and Inutsuka (2016), Copyright (2015), with permission from Elsevier

	$P > 0, d_{\rm dim} = 1$	$P < 0, d_{\rm dim} = 1$	$P > 0, d_{\rm dim} = 2$ or 3	$P < 0, d_{\rm dim} = 2$ or 3
Linear	◯	×	◯	×
Cubic	◯	×	×	◯
Quintic	×	◯	×	◯

$$V_{ij}^2 = \begin{cases} V_{ij,\rm cubic}^2 & \text{if } (P_i + P_j) > 0 \\ V_{ij,\rm quintic}^2 & \text{if } (P_i + P_j) < 0 \end{cases}, \tag{2.86}$$

while V_{ij}^2 in two and three dimensions is

$$V_{ij}^2 = \begin{cases} V_{ij,\rm linear}^2 & \text{if } (P_i + P_j) > 0 \\ V_{ij,\rm cubic}^2 & \text{if } (P_i + P_j) < 0 \end{cases}. \tag{2.87}$$

We also utilize the corresponding s_{ij}^* for each case.

Therefore, in cases of usual simulation setups, i.e., three-dimensional cases, we can conduct numerical simulations of elastic dynamics stably if we use the linear interpolation in regions with positive pressure and the cubic spline interpolation in those with negative pressure.

If the cubic or quintic spline interpolations are used when distance between two particles is much smaller than the smoothing length, V_{ij}^2 may diverge due to the interpolation. However, V_{ij}^2 should be about $1/\rho^2(\boldsymbol{x})$. Thus if V_{ij}^2 obtained from the cubic or quintic spline interpolation is much larger than $1/\rho^2(\boldsymbol{x})$, we should use linear interpolation. We suggest that the linear interpolation should be used when V_{ij}^2 becomes larger than $V_{ij,\rm crit}^2$ written as

$$V_{ij,\rm crit}^2 = 10\left(\frac{1}{\rho_{ij}^2}\right), \tag{2.88}$$

where $\rho_{ij} = (\rho_i + \rho_j)/2$.

2.1.5.4 Test Simulation 1: Elastic Rings Collision

To validate the above method to suppress the tensile instability using the Godunov SPH method, we conduct two test simulations. Firstly, we introduce a test simulation that reproduces collision of two elastic rings in a two-dimensional configuration. We use the following unit system: The density is scaled using undeformed density ρ_0,

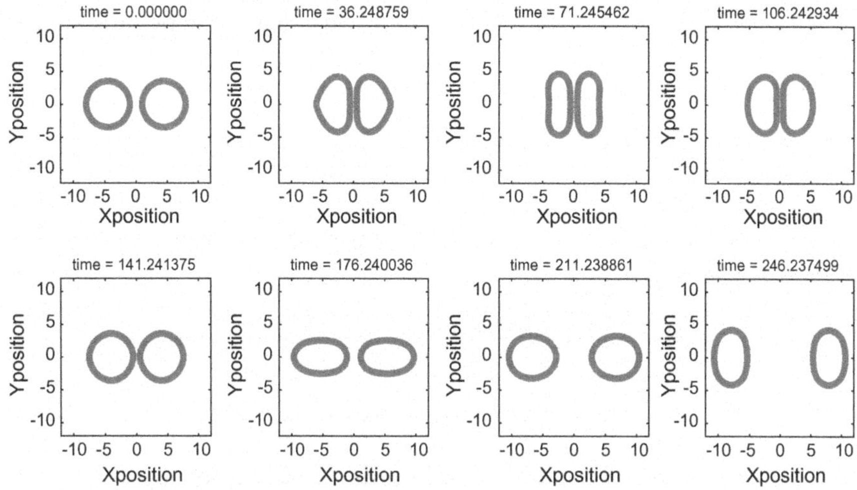

Fig. 2.4 Rubber rings collision calculated by the Godunov SPH method. Reprinted from Sugiura and Inutsuka (2017), Copyright (2016), with permission from Elsevier

the velocity is scaled using the bulk sound speed C_s, and the length is scaled using the widths of rings w. Initially the two rings are put with the distance between the centers of the two rings of $9w$. The rings have the inner radii of $3w$ and the outer radii of $4w$. We put particles on square lattices within the two rings with the smoothing length of $h = 0.1w$. The shear modulus is set to $\mu = 0.22C_s^2\rho_0$. The rings are not deformed initially, so that densities of all SPH particles are set to undeformed density ρ_0, and deviatoric stress tensors of all particles are set to 0. These rings collide with the relative velocity of $0.118C_s$.

Figure 2.4 shows the behavior of the elastic rings calculated using the Godunov SPH method, which validates the stability of the Godunov SPH method for the simulation of bounce off of the two rings. The configurations of the colliding rings are similar to those of Gray et al. (2001).

2.1.5.5 Test Simulation 2: Elastic Plate Oscillation

Secondly, to validate the Godunov SPH method in three-dimensional configurations, we simulate oscillation of an elastic plate with fixed end. The similar simulation but in two dimensions was conducted by Gray et al. (2001). Lifshitz et al. (1986) provide the analytical solution for extremely-thin-plate oscillations.

The same unit system as in the previous subsection is used except for the unit of length, which is scaled by the thickness of the plate H. The plate has the length of $L = 11H$ and the width of $W = 2H$. We put particles on square lattices within the plate with $h = 0.1w$. μ is set to $0.5C_s^2\rho_0$. Positions of particles that exist within $1H$ from the left end of the plate are fixed to represent fixed end of the plate. As in the previous subsection, we set initial densities to ρ_0 and initial deviatoric stress tensors to 0.

y component of the velocity v_y at the initial condition is given as

$$\frac{v_y}{C_s} = V_f \frac{[M(\cos(kx) - \cosh(kx)) - N(\sin(kx) - \sinh(kx))]}{Q}, \tag{2.89}$$

where x is the x position, V_f is the velocity normalized by the bulk sound speed at the free edge of the plate,

$$\begin{aligned} M &= \sin(kL) + \sinh(kL), \\ N &= \cos(kL) + \cosh(kL), \\ Q &= 2(\cos(kL)\sinh(kL) - \sin(kL)\cosh(kL)), \end{aligned} \tag{2.90}$$

and k is a kind of the wave number, which is given by the solution of

$$\cos(kL)\cosh(kL) = -1. \tag{2.91}$$

$kL = 1.875$ corresponds to a fundamental mode. In this simulation, we set $V_f = 0.05$.

Figure 2.5 shows the result of the simulation using the Godunov SPH method, which shows that our method calculates the oscillation stably.

According to Lifshitz et al. (1986), the angular frequency (=2π/oscillation period) of an extremely thin plate is written as

$$\omega^2 = \frac{EH^2k^4}{12\rho(1-\nu^2)}, \tag{2.92}$$

where E represents the Young's modulus and ν represents the Poisson's ratio, which are expressed using the bulk modulus K and the shear modulus μ as

$$\begin{aligned} E &= \frac{9K\mu}{3K+\mu}, \\ \nu &= \frac{3K-2\mu}{2(3K+\mu)}. \end{aligned} \tag{2.93}$$

The angular frequency using the parameters utilized in our simulation then becomes $\omega = 0.01201 C_s/H$, which leads to the analytical period of the oscillation of the extremely thin plate

$$T_{\rm theo} = \frac{2\pi}{\omega} = 523.2H/C_s. \tag{2.94}$$

Our simulation gives the period of $T_{\rm sim} \approx 665H/C_s$. Since the analytic period represents the period for extremely thin plates, the difference between the analytical and simulated periods is probably caused by the finite thickness of the simulated plate.

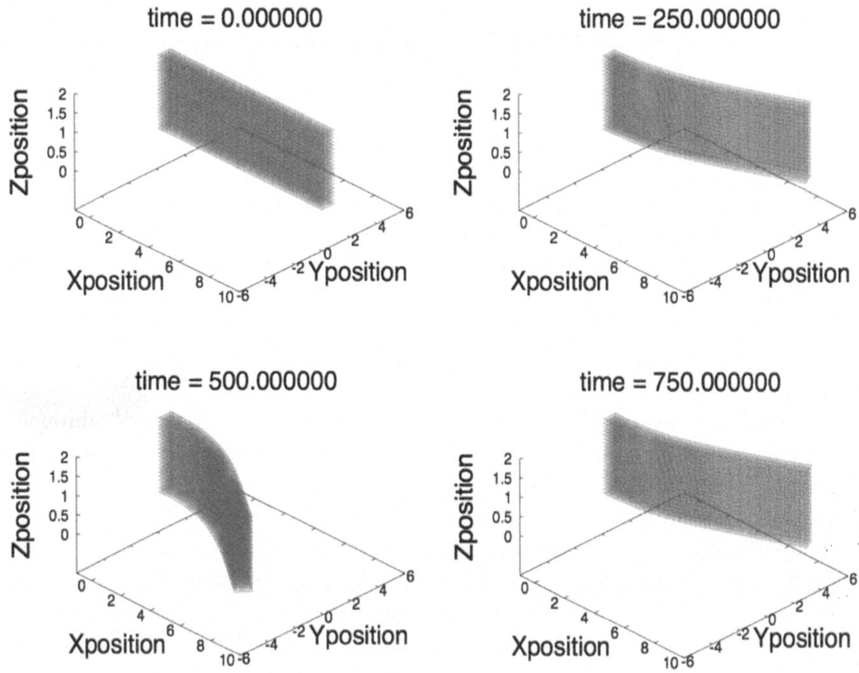

Fig. 2.5 Oscillation of a plate calculated by the Godunov SPH method. Reprinted from Sugiura and Inutsuka (2017), Copyright (2016), with permission from Elsevier

Thus the difference is expected to decrease with decreasing the ratio of the width to length of the plate W/L. We also conduct the same simulations but with $L = 15H$ and $L = 20H$. The oscillation periods with $L = 15H$ and $L = 20H$ are $1406H/C_s$ and $2353H/C_s$, respectively. The errors between simulated and theoretical results $(=(T_{\rm theo} - T_{\rm sim})/T_{\rm theo})$ are 27.3% for $L = 10H$, 19.5% for $L = 15H$, and 12.5% for $L = 20H$, respectively: The errors significantly decrease with decreasing W/L.

2.1.6 Time Development Method

Now, we have the time development equations for the velocity (2.20) or (2.65), the density (2.19), the specific internal energy (2.21) or (2.67), and the deviatoric stress tensor (2.23) or (2.69). There are several time advancing schemes such as Euler, Runge-Kutta, and leapfrog integrator. In this dissertation, we utilize the leapfrog integrator shown in Sugiura et al. (2018), which has the second-order accuracy in time.

The detailed procedures to update quantities of the ith particle at the nth step with the leapfrog integrator are as follows: The position $\boldsymbol{x}_i^n$ and other quantities q_i^n are updated as

$$\boldsymbol{x}_i^{n+1} = \boldsymbol{x}_i^n + \boldsymbol{v}_i^n \Delta t + \frac{1}{2}\left(\frac{d\boldsymbol{v}_i}{dt}\right)^n \Delta t^2, \tag{2.95}$$

$$q_i^{n+1} = q_i^n + \frac{1}{2}\left[\left(\frac{dq_i}{dt}\right)^n + \left(\frac{dq_i}{dt}\right)^{n+1}\right]\Delta t, \tag{2.96}$$

where Δt is the time step and $q = \rho, u, \boldsymbol{v}, S^{\alpha\beta}$. Here, we need $(dq_i/dt)^{n+1}$ to determine q_i^{n+1}. However, the determination of $(dq_i/dt)^{n+1}$ requires q_i^{n+1} (see, e.g., the equation of motion (2.20)). Thus, we firstly predict the quantities at the $(n+1)$th step as

$$q_i^* = q_i^n + \left(\frac{dq_i}{dt}\right)^n \Delta t. \tag{2.97}$$

We also update the positions as in Eq. (2.95). We then calculate $(dq_i/dt)^{n+1}$ through, e.g., the equation of motion, using q_i^* and the updated positions $\boldsymbol{x}_i^{n+1}$. We finally obtain q_i^{n+1} as shown in Eq. (2.96).

We do not need to calculate both $(dq_i/dt)^n$ and $(dq_i/dt)^{n+1}$ at every step since we again utilize $(dq_i/dt)^{n+1}$ as $(dq_i/dt)^n$ at next step, which means that we need to calculate the time derivatives through, e.g., the equation of motion, only once at each step. In spite of the low computational cost of the leapfrog integrator, the scheme provides the second order accuracy in time, i.e., the scheme does not cause integration errors if all variables show linear time variation.

According to the Courant condition, the time step Δt is given by

$$\Delta t = \min_i C_{\mathrm{CFL}} \frac{h}{C_{s,i}}, \tag{2.98}$$

where C_{CFL} is a parameter that should be smaller than unity to satisfy the Courant condition, and $C_{s,i}$ is the local bulk sound speed at the position of the ith particle. We set $C_{\mathrm{CFL}} = 0.5$. Bulk sound speed for elastic dynamics is defined as in Eq. (2.4). Even for other equation of states, $C_{s,i}$ is expressed as $C_{s,i} = \sqrt{dP_i/d\rho_i}$ and is calculated using ρ, u, and P of the ith particle.

2.2 Tillotson Equation of State

For elastic dynamics, we use the simple equation of state Eq. (2.4) where the pressure is proportional to the density. However, Eq. (2.4) cannot treat the states caused by hypervelocity impacts such as melting, vaporization, and extreme pressure produced by shock waves.

The Tillotson equation of state is one of the most frequently used equation of states for numerical simulations of celestial body impacts (e.g., Genda et al. 2012; Jutzi and Asphaug 2015; Leleu et al. 2018). Tillotson (1962) developed the Tillotson equation of state for metallic materials through laboratory experiments of impacts.

For low-energy and low-deformation states, the Tillotson equation of state converges to elastic equation of state Eq. (2.4), and for high-energy states, i.e., vaporized states, this converges to the ideal-gas equation of state. Thus the Tillotson equation of state can treat both low-energy elastic solids and high-energy vaporized gas. The Tillotson equation of state has ten material dependent parameters. Some previous studies determined these material dependent parameters for basaltic and icy materials (Benz and Asphaug 1999; Lange and Ahrens 1983; Nakamura and Fujiwara 1991; O'Keefe and Ahrens 1982a, b).

The Tillotson equation of state has three different forms depending on values of the density and the specific internal energy. For $\rho > \rho_0$ or $u < u_{\rm iv}$, where ρ_0 is the uncompressed density and $u_{\rm iv}$ is the specific internal energy for incipient vaporization, the equation is

$$P = P_s(\rho, u) = \left[a_{\rm til} + \frac{b_{\rm til}}{(u/(u_{0,\rm til}\eta^2)+1)}\right]\rho u + A_{\rm til}\mu + B_{\rm til}\mu^2, \tag{2.99}$$

where $\eta = \rho/\rho_0$ and $\mu = \eta - 1$. Equation (2.99) is for compressed or cold expanded states. For $\rho < \rho_0$ and $u > u_{\rm cv}$, where $u_{\rm cv}$ is the specific internal energy for complete vaporization, the equation is

$$P = P_g(\rho, u) = a_{\rm til}\rho u + \left[\frac{b_{\rm til}\rho u}{(u/(u_{0,\rm til}\eta^2)+1)} + A_{\rm til}\mu e^{-\beta_{\rm til}(\rho_0/\rho - 1)}\right]e^{-\alpha_{\rm til}(\rho_0/\rho-1)^2}. \tag{2.100}$$

Equation (2.100) is for expanded and completely vaporized states. Finally, for $\rho < \rho_0$ and $u_{\rm iv} < u < u_{\rm cv}$, the equation is

$$P = \frac{(u - u_{\rm iv})P_g(\rho, u_{\rm cv}) + (u_{\rm cv} - u)P_s(\rho, u_{\rm iv})}{u_{\rm cv} - u_{\rm iv}}. \tag{2.101}$$

Equation (2.101) is for expanded and partially vaporized states. Here, ρ_0, $u_{\rm iv}$, $u_{\rm cv}$, $u_{0,\rm til}$, $A_{\rm til}$, $B_{\rm til}$, $a_{\rm til}$, $b_{\rm til}$, $\alpha_{\rm til}$, and $\beta_{\rm til}$ are material dependent Tillotson parameters.

Equation (2.99) converges to the elastic equation of state Eq. (2.4) if $u \to 0$ and $\rho \approx \rho_0$. We also notice that $A_{\rm til}$ corresponds to the bulk modulus for low-temperature and nearly-uncompressed states, and thus the bulk sound speed of the Tillotson equation of state for such states is calculated as

$$C_{s,\rm til} = \sqrt{\frac{A_{\rm til}}{\rho_0}}. \tag{2.102}$$

Equation (2.100) converges to the ideal gas equation of state if $u \to \infty$, and $\alpha_{\rm til}$ and $\beta_{\rm til}$ control the speed of the convergence to the ideal gas equation of state.

2.3 Self Gravity

2.3.1 *Self Gravity for SPH Method*

The SPH method uses particles that mimic small fractions of fluid or elastic bodies, and each SPH particle has its own mass. Thus, as in the case of N-body orbital simulations, the self-gravity between the ith and jth SPH particles is directly calculated as

$$g_{ij,\mathrm{simple}} = -Gm_j \frac{\boldsymbol{x}_i - \boldsymbol{x}_j}{|\boldsymbol{x}_i - \boldsymbol{x}_j|^3}, \tag{2.103}$$

where G is the gravitational constant.

However, Eq. (2.103) is not correct especially for close pairs of particles, because Eq. (2.103) is derived using the assumption that all particles are point masses. As we introduced in Sect. 2.1.2, SPH particles are not point masses but have a kind of radii. We cannot ignore the broadening of SPH particles especially for the calculation of the self gravity between close pairs of particles. Crucially, Eq. (2.103) diverges if two particles are completely overlapped, but this does not occur if we correctly treat the broadening of SPH particles.

Here, for the calculation of the self gravity between a pair of SPH particles, we only consider the broadening of one particle of the pair. This seems to break the action-reaction principle, which leads to nonconservation of the momentum, but our simulations do not have such problem since we use the same broadening radius, or the smoothing length, for all SPH particles. Then Eq. (2.103) is modified as

$$\begin{aligned} g_{ij} &= -G\hat{m}_j \frac{\boldsymbol{x}_i - \boldsymbol{x}_j}{|\boldsymbol{x}_i - \boldsymbol{x}_j|^3}, \\ \hat{m}_j &= \int_0^{|\boldsymbol{x}_i - \boldsymbol{x}_j|} 4\pi r^2 m_j W(r, h) dr. \end{aligned} \tag{2.104}$$

Eq. (2.104) converges to zero if $|\boldsymbol{x}_i - \boldsymbol{x}_j| \to 0$.

The detailed expression of $\hat{m}_j$ in the case of the Gaussian kernel is

$$\frac{\hat{m}_j}{m_j} = \left[\mathrm{erf}\left(\frac{|\boldsymbol{x}_i - \boldsymbol{x}_j|}{h}\right) - \frac{2|\boldsymbol{x}_i - \boldsymbol{x}_j|}{h\sqrt{\pi}} \exp\left(-\frac{|\boldsymbol{x}_i - \boldsymbol{x}_j|^2}{h^2}\right)\right], \tag{2.105}$$

where erf is the error function, while that in the case of the cubic spline kernel is

$$\frac{\hat{m}_j}{m_j} = \begin{cases} \frac{4}{3}(\Delta_{ij})^3 - \frac{6}{5}(\Delta_{ij})^4 + \frac{1}{2}(\Delta_{ij})^5 & 0 \le \Delta_{ij} < 1 \\ \frac{8}{3}(\Delta_{ij})^3 - 3(\Delta_{ij})^4 + \frac{6}{5}(\Delta_{ij})^5 - \frac{1}{6}(\Delta_{ij})^6 - \frac{1}{15} & 1 \le \Delta_{ij} < 2 \\ 1 & 2 \le \Delta_{ij} \end{cases}, \tag{2.106}$$

where $\Delta_{ij} = |\boldsymbol{x}_i - \boldsymbol{x}_j|/h$.

If we want to calculate the self gravity, the equation of motion Eq. (2.20) is modified as

$$\frac{dv_i^\alpha}{dt} = \sum_j m_j \left[\frac{\sigma_i^{\alpha\beta}}{\rho_i^2} + \frac{\sigma_j^{\alpha\beta}}{\rho_j^2} - \Pi_{ij}\delta^{\alpha\beta}\right] \frac{\partial}{\partial x_i^\beta} W(|\boldsymbol{x}_i - \boldsymbol{x}_j|, h) + \sum_j g_{ij}^\alpha. \quad (2.107)$$

2.3.2 Acceleration Scheme of Computation of Self Gravity

If we directly compute the self gravity between all pairs of SPH particles, the computational cost increases with $\mathcal{O}(N^2)$, where N is the total number of SPH particles. Thus the computational cost is proportional to N^2, which means that the direct computation of the self gravity is difficult especially for large number of particles (typically more than 100,000).

Barnes and Hut (1986) developed a novel technique that utilizes a tree structured subdivision of space, which enables us to reduce the computational cost to $\mathcal{O}(N \log N)$. In this technique, firstly we divide whole space that includes all SPH particles (root cell) into eight subcells, and then we recursively divide each subcell into eight subcells whenever more than several particles occupy the same cell. Then we calculate the total mass and the center of mass of SPH particles that are included in each subcell. Finally we calculate the self gravity on an SPH particle from all other particles as the following recursive procedure:

- We assess whether a subcell is sufficiently faraway from the particle or not with inequality $l/d < \theta$, where l is the side length of this subcell, d is the distance from the particle to the center of mass of this subcell, and θ determines the accuracy and has $0.3 - 1.0$ (A).

- If yes, we calculate the self gravity on the particle from the center of mass of this subcell as the representative gravity from all particles in this subcell.
- If no, we do the same assessment (A) for direct subcells of this subcell.

The assessment (A) is firstly done for the root cell. If leaf cells do not satisfy the condition $l/d < \theta$, then we directly calculate the gravity from the particles in leaf cells. Following this procedure, we can reduce the computational cost to $\mathcal{O}(N \log N)$.

2.4 Fracture Model

In Sect. 2.1, we introduced how to simulate the dynamics of elastic bodies. Asteroids are certainly elastic bodies if deformation is small. However, asteroids are rocky material, which is fractured under large deformation. Thus we need an appropriate model for the fracture to describe collisional destruction. In this section, we explain the fracture model for the SPH method introduced by Benz and Asphaug (1995), who follow the fracture model of brittle solids developed by Grady and Kipp (1980).

2.4.1 Concept of Fracture Model

Rock is fractured under large tensile forces. Rocks do not have uniform strength, but there are some weak parts and also some strong parts. Fractures are typically generated at weak parts and extend to other parts.

Benz and Asphaug (1995) assume the existence of explicit incipient flaws in rocky material. The probability distribution of the strength of incipient flaws is expressed as Weibull distribution (Weibull 1939). The probable number of flaws per unit volume that have the failure strain smaller than ϵ is expressed as a power law form

$$n(\epsilon) = k_w \epsilon^{m_w}, \tag{2.108}$$

where k_w and m_w are Weibull parameters, which can be determined from laboratory experiments. The parameter k_w determines the number of flaws per unit volume and widely varies between various rock types. The parameter m_w determines the dispersion of strengths; small m_w means rocks with widely varying strengths, while large m_w means homogeneous rocks. m_w typically has values from 6 to 12. Note that the Weibull distribution itself does not have the information about the position or direction of flaws. Using Eq. (2.108), we can also derive a flaw's activation threshold of strain ϵ for a flaw.

Here, we consider a part of a rock. When local tensile strain has exceeded a threshold of strain for a flaw, a crack is generated and allowed to grow. According to Lawn (1993), the speed of crack extension is a constant C_g, which is about 0.4 times the speed of a longitudinal sound wave. Thus the length of the growing crack l_c is expressed as

$$l_c = C_g(t - t_0), \tag{2.109}$$

where t_0 is the time of crack generation. Grady and Kipp (1980) introduced a damage parameter D that has the range of $0 - 1$. $D = 0$ represents an intact rock, while $D = 1$ represents a completely damaged rock that does not feel any tensile stress. A rock with $D = 0.5$ feels only half of the original tensile stress calculated from elastic dynamics equations. Benz and Asphaug (1995) defined D as the local fraction of volume that has relieved stress by cracks. According to Walsh (1965), the volume that has relieved stress by a crack approximately equals to the volume of the sphere with the radius of the crack length. Thus D is expressed as

$$D = \frac{(4/3)\pi l_c^3}{V_s}, \tag{2.110}$$

where $V_s = (4/3)\pi R_s^3$ is the volume of the part of the rock that we consider now. Using Eqs. (2.109) and (2.110), we can derive the differential equation for D as

$$\frac{dD^{1/3}}{dt} = \frac{C_g}{R_s}. \tag{2.111}$$

Tensile force, that is, negative pressure is relieved by fractures. According to the definition of D, the pressure is modified as

$$P_D = \begin{cases} (1-D)P & P < 0 \\ P & P > 0 \end{cases}. \tag{2.112}$$

P_D is used in the equation of motion and so on as the pressure. Shear force, that is, the deviatoric stress is also affected by fractures. However, the deviatoric stress in a part with $D = 1$ is mainly determined by the friction between completely destructed rock. Thus we leave the explanation of the treatment of the deviatoric stress for damaged rocks to Sect. 2.5.

2.4.2 SPH Implementation of Fracture Model

According to Benz and Asphaug (1995), we explicitly assign threshold strains for crack activation to each SPH particle when we prepare initial conditions. We consider an asteroid with total volume of V. The weakest flaw has the threshold strain of

$$\epsilon_1^{\text{act}} = \left[\frac{1}{k_w V}\right]^{1/m_w}. \tag{2.113}$$

In the same manner, the pth weakest flaw has the threshold strain of

$$\epsilon_p^{\text{act}} = \left[\frac{p}{k_w V}\right]^{1/m_w}. \tag{2.114}$$

We randomly choose an SPH particle and assign a threshold ϵ_p^{act} with $p = 1$ to this particle, and next we choose another SPH particle randomly and assign a threshold ϵ_p^{act} with $p = 2$, and next we choose another one and assign that with $p = 3 \ldots$ We continue this procedure until all SPH particles have at least one activation threshold. Each SPH particle has, on average, $\ln N$ flaws, where N is the number of SPH particles. Hereafter, we represent the threshold strain for the pth weakest flaw assigned to the ith SPH particle as $\epsilon_{i,p}^{\text{act}}$.

According to Melosh et al. (1992) or Benz and Asphaug (1995), local strain at the ith SPH particle is calculated from damage-relieved stress tensor $\sigma_{D,i}^{\alpha\beta} = -P_{D,i}\delta^{\alpha\beta} + S_{D,i}^{\alpha\beta}$, where $S_{D,i}^{\alpha\beta}$ is damage-relieved deviatoric stress tensor and will be introduced in Sect. 2.5. We firstly calculate three principal stresses $\sigma_{D,i}^1$, $\sigma_{D,i}^2$, and $\sigma_{D,i}^3$, which are eigenvalues of the damage-relieved stress tensor. Then we find the maximum principal stress $\sigma_{D,i}^{\mathrm{m}}$, which is the maximum tensile stress among the three principal stresses. The local strain at the ith SPH particle ϵ_i is calculate as

$$\epsilon_i = \frac{\sigma_{D,i}^{\mathrm{m}}}{(1 - D_i)E}, \tag{2.115}$$

where $E = 9K\mu/(3K + \mu)$ is the Young's modulus.

If at least one of the flaws inside of the ith particle is activated, i.e, at least one of $\epsilon_{i,p}^{\mathrm{act}}$ is smaller than ϵ_i, the damage parameter D_i increases according to

$$\frac{dD_i^{1/3}}{dt} = \frac{C_g}{R_s}. \tag{2.116}$$

Here, $R_s = 4h$ for the Gaussian kernel and $R_s = 2h$ for the cubic spline kernel. However, D_i is only allowed to accumulate to a maximum value given by

$$D_i^{\max} = \left(\frac{n_i}{n_i^{\mathrm{tot}}}\right)^{1/3}, \tag{2.117}$$

where n_i^{tot} is the total number of flaws inside of the ith particle, and n_i is the number of activated flaws inside of the ith particle. If multiple flaws are activated, the damage parameter will increase faster, and thus we multiply n_i to the right hand side of Eq. (2.116). According to Eq. (2.112), the pressure of the ith particle is modified as

$$P_{D,i} = \begin{cases} (1 - D_i)P_i & P_i < 0 \\ P_i & P_i > 0 \end{cases}. \tag{2.118}$$

2.5 Pressure Dependent Failure Model and Friction Model

Benz and Asphaug (1995) set $S_{D,i}^{\alpha\beta} = (1 - D)S_i^{\alpha\beta}$ for damaged SPH particles. This means that completely damaged SPH particles with $D = 1$ behave like fluid that does not feel any tensile and shear stresses. However, completely damaged rocks, or granular material, have the friction between granular particles. The frictional forces are shear forces and play the crucial role for the shape formation of rubble pile bodies. Jutzi (2015) introduced the friction model to the SPH method, and he also introduced the pressure dependent failure model for intact rocks. In this section, we introduce these models. We also show results of test simulations using our SPH code with the friction model, since the friction is the most important effect for the formation of shapes of rubble pile bodies.

2.5.1 Description of Models

Actual metals are not perfect elastic bodies but plastic deformation occurs under large deformation. It is called as yielding, and the criterion of yielding for metallic material is described by the pressure independent von Mises yield criterion. However, yielding criterion of rocky material is pressure dependent, and shear strength of intact rocks increases with increasing confining pressure. According to Collins et al. (2004) or Jutzi (2015), the pressure dependent yielding strength for intact rock at the ith SPH particle Y_i is expressed as

$$\begin{aligned} Y_i = Y_{I,i} &= \left(Y_0 + \frac{\mu_I P_{\mathrm{pos},i}}{1 + \mu_I P_{\mathrm{pos},i}/(Y_M - Y_0)} \right) \left(1 - \frac{u_i'}{u_{\mathrm{melt}}} \right), \\ P_{\mathrm{pos},i} &= \max[P_i, 0], \\ u_i' &= \min[u_i, u_{\mathrm{melt}}], \end{aligned} \tag{2.119}$$

where Y_0 and Y_M are the yielding strength at zero pressure and infinite pressure, respectively, μ_I is the coefficient of internal friction, and u_{melt} is the specific internal energy for melting.

For completely damaged rocks, the yielding strength is determined by the friction. As we know, the friction is proportional to confining pressure. Thus the yielding strength for the completely damaged rock at the ith SPH particle is expressed as

$$Y_i = Y_{D,i} = \mu_d P_{i,\mathrm{pos}}, \tag{2.120}$$

where μ_d is the friction coefficient, which is the most important parameter for our study. The angle of friction is defined as $\phi_d = \arctan(\mu_d)$ and generally the same as the angle of repose. Note that μ_d here merely determines the ratio of frictional stress to confining pressure.

For partially damaged rocks, we use the following yielding strength that allows a smooth transition between Eqs. (2.119) and (2.120) (Jutzi 2015):

$$Y_i = (D_i - 1) Y_{I,i} + D_i Y_{D,i}. \tag{2.121}$$

Here, we use the following limitations $Y_{D,i} \leq Y_{I,i}$ and $Y_i \leq Y_{I,i}$.

To implement the yielding, we usually utilize the second invariant of the deviatoric stress tensor $J_{2,i}$, which is computed as

$$J_{2,i} = \frac{1}{2} S_i^{\alpha\beta} S_i^{\alpha\beta}. \tag{2.122}$$

$\sqrt{J_{2,i}}$ is a kind of magnitude of the deviatoric stress tensor $S_i^{\alpha\beta}$, and $S_i^{\alpha\beta}/\sqrt{J_{2,i}}$ represents a kind of unit vector of this deviatoric stress tensor. Thus $Y_i S_i^{\alpha\beta}/\sqrt{J_{2,i}}$ has the magnitude of Y_i and the direction of $S_i^{\alpha\beta}$; $Y_i S_i^{\alpha\beta}/\sqrt{J_{2,i}}$ represents the shear stress under the yielding. If the magnitude $\sqrt{J_{2,i}}$ is larger than Y_i, the yielding occurs and the deviatoric stress tensor should be modified to $Y_i S_i^{\alpha\beta}/\sqrt{J_{2,i}}$. Thus we modify the deviatoric stress tensor as

$$S_{D,i}^{\alpha\beta} = f S_i^{\alpha\beta}, \tag{2.123}$$
$$f = \min\left[\frac{Y_i}{\sqrt{J_{2,i}}}, 1\right].$$

In each time step, we also modify $S_i^{\alpha\beta}$ as $S_i^{\alpha\beta} = S_{D,i}^{\alpha\beta}$.

2.5.2 Test Simulations for Friction Model

To confirm the validity of the friction model implemented in our simulation code, we conduct test simulations that reproduce cliff collapse experiments following the test simulations in Jutzi (2015). Lajeunesse et al. (2005) conducted the laboratory experiments of cliff collapse of glass beads in a two-dimensional configuration. They conducted the experiments with grass beads with diameters of 1.15 mm and 3 mm, various initial height of cliff H_0, and various initial ratios of length of cliff L_0 to height of cliff H_0. As a result, they found that final configurations only depend on the friction angle of beads and the initial aspect ratio L_0/H_0. They also found that profiles of cliffs at different times and different experiments are barely distinguishable when the profiles are scaled by L_0 and the times are scaled by the characteristic time $\tau = \sqrt{H_0/g}$, where g is the gravitational acceleration on surfaces of the Earth.

For the simulation of the cliff collapse, we use the standard SPH method and the cubic spline kernel. Damage parameters of all SPH particles are initially set to unity, and thus SPH particles mimic granular material. The friction coefficient is set to $\mu_d = \tan(20.9°)$, and compare with the experiments using beads with the same friction angle. We use the simple equation of state Eq. (2.4) with $C_s = 500\,\mathrm{m/s}$ and $\rho_0 = 1500\,\mathrm{kg/m^3}$, which are the typical values for granular material. We do not calculate the self gravity, but add the downward gravitational acceleration g to the equation of motion of SPH particles. To mimic the wall and floor in the experiments, we put several layers of SPH particles with fixed positions on just outside of the simulation space. We set $L_0 = 1,000\,\mathrm{m}$ to avoid too small time steps (see Eq. 2.98). Although the length of cliff is much larger than the experimental scale, this is not the problem because results can be scaled by L_0 and τ.

Figure 2.6 shows the comparison between our simulations, the simulations of Jutzi (2015), and the experiments of Lajeunesse et al. (2005), and we notice that these results agree well with each other. Therefore, our simulation code can well reproduce behaviors of granular bodies.

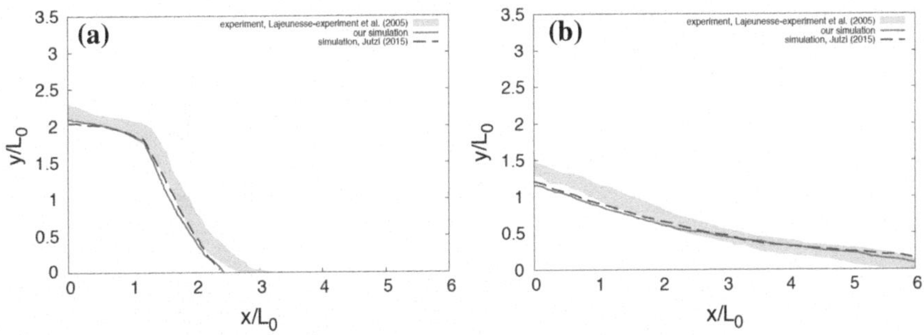

Fig. 2.6 Profiles of collapsing cliff with $H_0/L_0 = 3.2$ at **a** $t = \tau$ and **b** $t = \infty$. The red solid curves show the results of our simulations, the blue dashed curves show the results of the simulations of Jutzi (2015), and the cyan shaded regions show the results obtained from the experiments of Lajeunesse et al. (2005). The results of the experiments are shown in shaded regions because we describe the profiles obtained from several experiments with the same L_0/H_0 and the friction angle, which have small scatters

2.6 Parallelization of Simulation Code Using FDPS

In this dissertation, we reproduce three-dimensional shapes of asteroids through numerical simulations using the SPH method. Three-dimensional simulations require larger number of SPH particles compared to one- or two-dimensional simulations. Especially, to investigate shapes of smaller remnants produced through disruptive impacts, we need much larger number of SPH particles of about several millions. We cannot complete such high-resolution simulations with a normal laptop PC within acceptable computational time, and we need to utilize supercomputers. This means that we have to parallelize our simulation code to utilize many CPU cores.

Usual supercomputers are composed of PC clusters with distributed memories. Thus for efficient parallelization, we have to consider amounts of sending/receiving data and load balancing between CPU cores. If the amounts of sending/receiving data are too large or workload balance is not good, parallelization efficiency becomes significantly low, which means that it is meaningless to utilize many CPU cores. However, in general, it is very difficult to parallelize particle-based simulation codes so that they achieve high parallelization efficiency, because particles move from one CPU domain to the other. Thus we need to reassign SPH particles to other CPU cores so as to keep good workload balance. Moreover, the simulation codes that calculate the self gravity using the tree method (see Sect. 2.3.2) are so complex that usual researchers cannot parallelize such codes with acceptable efforts.

To support efficient parallelization of codes based on particle methods, a particle simulator research team, AICS, RIKEN developed Framework for Developing Particle Simulator (FDPS: Iwasawa et al. 2015, 2016). Load balancing among CPUs is important for the efficient parallelization, and FDPS provides a framework that enables particles' information exchange between CPUs and neighbor particle searching with good load balancing. FDPS also provides a function to calculate the self

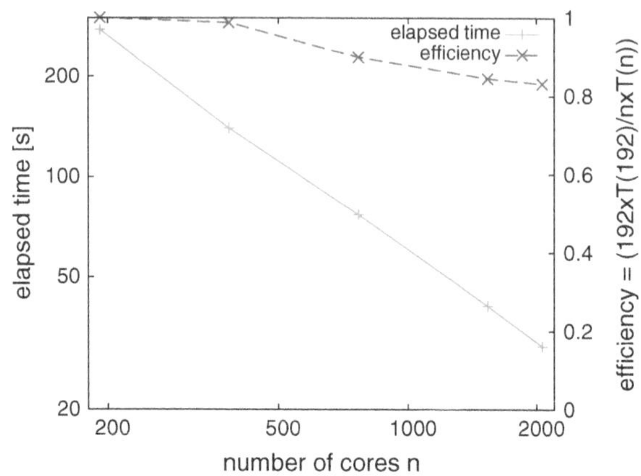

Fig. 2.7 Parallelization efficiency of our simulation code. The red solid line shows elapsed times and the blue dashed line shows parallelization efficiency defined as Eq. (2.124). The horizontal axis represents the number of utilized CPU cores, the left vertical axis represents the elapsed time, and the right vertical axis represents the parallelization efficiency

gravity using the tree method shown in Sect. 2.3.2. Our simulation code is parallelized by making use of FDPS, and thus we can efficiently utilize parallel computers for our simulations.

We checked the parallelization efficiency of our simulation code. We calculated the parallelization efficiency through the strong scaling, i.e., we measured the elapsed times needed to simulate the problem with fixed number of SPH particles but with various number of CPU cores. We used the standard SPH method with the cubic spline kernel. We used about 10^8 SPH particles and solved an elastic problem for 10 time steps. We utilized XC30 system of Center for Computational Astrophysics, National Astronomical Observatory of Japan.

We vary utilized CPU cores from 192 to 2,064. The parallelization efficiency is calculated based on the elapsed time when we use 192 cores as

$$\text{efficiency} = 192T(192)/(nT(n)), \tag{2.124}$$

where n is the number of utilized CPU cores and $T(n)$ shows the elapsed time to simulate the elastic problem when we use n CPU cores. From Fig. 2.7, we notice that the elapsed time decreases linearly with increasing the number of utilized CPU cores. We also notice that the parallelization efficiency when we use 2,064 CPU cores is about 0.8, which is the quite high efficiency. This means that the elapsed time with 2064 CPU cores is about nine times shorter than that with 192 CPU cores. Thus, our simulation code is well efficiently parallelized thanks to FDPS.

2.7 Summary of Utilized Methods and Parameters

The standard SPH method cannot treat elastic dynamics problems stably due to the tensile instability, but it only needs low computational costs. The tensile instability is insignificant for asteroidal collisions because fractures of rocky material quickly release negative pressure, and we can still utilize the standard SPH method for such problems. Thus hereafter in this dissertation, we utilize the standard SPH method that needs low computational costs to conduct detailed parameter surveys or high-resolution simulations. We also utilize the following methods or models: the leap-frog integrator (Sect. 2.1.6), the correction matrix for rigid body rotations (Sect. 2.1.3), the Tillotson equation of state (Sect. 2.2), the self gravity (Sect. 2.3), the fracture model (Sect. 2.4), the pressure dependent failure model (Sect. 2.5), and the friction model (Sect. 2.5).

Throughout this dissertation, we assume the material of asteroids as basalt. There are many material dependent parameters required for the simulations of asteroidal collisions such as the shear modulus or Tillotson parameters. Here, we summarize the parameters utilized for the simulations in following sections. These parameters are summarized in Tables 2.2, 2.3, 2.4, 2.5. Note that we vary the friction angle in Chap. 5 and we do not list the value of the friction angle here.

Table 2.2 Parameters for elastic dynamics. μ shows the shear modulus and K shows the bulk modulus

μ(Pa)	K(Pa)
2.27×10^{10}	2.67×10^{10}

Table 2.3 Tillotson parameters for basalt. These values are listed in Benz and Asphaug (1999)

ρ_0(kg/m^3)	u_{iv}(J/kg)	u_{cv}(J/kg)	$u_{0,til}$(J/kg)	A_{til}(Pa)	B_{til}(Pa)	a_{til}	b_{til}	α_{til}	β_{til}
2.7×10^3	4.72×10^6	1.82×10^7	4.87×10^8	2.67×10^{10}	2.67×10^{10}	0.5	1.5	5.0	5.0

Table 2.4 Parameters for the fracture model, or Weibull parameters. These values are listed in Benz and Asphaug (1999)

k_w(m^{-3})	m_w
4.0×10^{35}	9.0

Table 2.5 Parameters for the pressure dependent failure model. The values for Y_0, Y_M, and μ_I are taken from Jutzi (2015). The value for u_{melt} is listed in Benz and Asphaug (1999)

Y_0(Pa)	Y_M(Pa)	μ_I	u_{melt}(J/kg)
1.0×10^8	3.5×10^9	1.5	3.4×10^6

Appendix: Linear Stability Analysis for Godunov SPH Method

Here, we introduce the details of the linear stability analysis of the equations for the Godunov SPH method. In this linear stability analysis, we use the following simplification. We ignore the terms of the deviatoric stress tensor and set $S_i^{\alpha\beta} = 0$ because the deviatoric stress tensor generally prevents the clumping of SPH particles, which leads to prevent the tensile instability. We also ignore the Riemann solver and set $P_{ij}^* = (P_i + P_j)/2$ because numerical viscosity terms including the Riemann solver generally prevent numerical instabilities. We consider the cases that SPH particles are put on cubic lattices, and the smoothing length h is set to be the side length of the cubic lattices Δx. We conduct the linear stability analysis for the longitudinal perturbations because the tensile instability is the instability of compressional waves. We utilize the equation of state of $P = C_s^2(\rho - \rho_0)$. Unperturbed states have the uniform density and pressure, and all SPH particles have the same mass m.

Unperturbed positions of particles are expressed as,

$$\overline{\boldsymbol{x}}_i = (\overline{x}_i, \overline{y}_i, \overline{z}_i). \tag{2.125}$$

We add the perturbation of positions in the x-direction. Then the positions of particles are represented as

$$\begin{aligned} \boldsymbol{x}_i &= (\overline{x}_i + \delta x_i, \overline{y}_i, \overline{z}_i), \\ \delta x_i &= \epsilon_x \exp[I(k\overline{x}_i - \omega t)], \end{aligned} \tag{2.126}$$

where ϵ_x is very small value, k and ω represent the wavenumber and frequency of the perturbation respectively, and I represents the imaginary unit. Hereafter, ϵ shows very small value, and we ignore second or higher order of ϵ.

From Eq. (2.126) and $(d/dt)\boldsymbol{x}_i = \boldsymbol{v}_i$, the velocity of the ith particle is represented as

$$\boldsymbol{v}_i = (-I\omega\delta x_i, 0, 0). \tag{2.127}$$

The density in unperturbed states is now defined as $\overline{\rho}$, and the density of the ith particle is written as

$$\begin{aligned} \rho_i &= \overline{\rho} + \delta\rho_i, \\ \delta\rho_i &= \epsilon_\rho \exp[I(k\overline{x}_i - \omega t)]. \end{aligned} \tag{2.128}$$

From Eq. (2.19), we represent $\delta\rho_i$ using δx_i as,

$$\begin{aligned} \delta\rho_i &= -I\overline{\rho}D\delta x_i, \\ D &\equiv \sum_j -\sin[k(\overline{x}_i - \overline{x}_j)]\frac{\partial}{\partial \overline{x}_i} W(|\overline{\boldsymbol{x}}_i - \overline{\boldsymbol{x}}_j|, h)\frac{m}{\overline{\rho}}. \end{aligned} \tag{2.129}$$

Then we represent the pressure of the ith particle as

$$P_i = \overline{P} + \delta P_i = \overline{P} + C_s^2 \delta\rho_i = \overline{P} - I C_s^2 \overline{\rho} D \delta x_i, \tag{2.130}$$

where $\overline{P} = C_s^2(\overline{\rho} - \rho_0)$ represents the pressure in unperturbed states.

From Eq. (2.54), the specific volume is represented as

$$V_i = \frac{1}{\overline{\rho}}(1 + I D \delta x_i). \tag{2.131}$$

From Eq. (2.54), the x-component of the specific volume gradient is represented as

$$\begin{aligned}
\frac{\partial V_i}{\partial x_i} &= -\frac{1}{\rho_i^2}\frac{\partial \rho_i}{\partial x_i} \\
&\approx -\frac{1}{\overline{\rho}^2}\sum_j m(\delta x_i - \delta x_j)\frac{\partial^2}{\partial \overline{x_i}^2} W(|\overline{\boldsymbol{x}_i} - \overline{\boldsymbol{x}_j}|, h) = -\frac{1}{\overline{\rho}} C_\rho \delta x_i, \\
C_\rho &\equiv \sum_j (1 - \cos[k(\overline{x_i} - \overline{x_j})])\frac{\partial^2}{\partial \overline{x_i}^2} W(|\overline{\boldsymbol{x}_i} - \overline{\boldsymbol{x}_j}|, h)\frac{m}{\overline{\rho}}.
\end{aligned} \tag{2.132}$$

The xx-component of the second derivative of V_i is represented using Eq. (2.54) as

$$\frac{\partial^2 V_i}{\partial x_i^2} = \sum_j \frac{m}{\rho_j}\frac{\partial V_j}{\partial x_j}\frac{\partial}{\partial x_i} W(|\boldsymbol{x}_i - \boldsymbol{x}_j|, h) \approx -\frac{I}{\overline{\rho}} C_\rho D \delta x_i. \tag{2.133}$$

We then substitute linearized quantities into Eq. (2.65), and finally we obtain the dispersion relations (DRs) for the three interpolation methods. DR for the linear interpolation is expressed as

$$\omega_{\text{linear}}^2 = -C_s^2 D a + \frac{\overline{P}}{\overline{\rho}}\left[2Da + 2b\right], \tag{2.134}$$

where

$$\begin{aligned}
a &= \sum_{j \neq i} \sin[k(\overline{x_i} - \overline{x_j})]\frac{\partial}{\partial \overline{x_i}} W(|\overline{\boldsymbol{x}_i} - \overline{\boldsymbol{x}_j}|, \sqrt{2}h)\frac{m}{\overline{\rho}}, \\
b &= \sum_{j \neq i} (1 - \cos[k(\overline{x_i} - \overline{x_j})])\frac{\partial^2}{\partial \overline{x_i}^2} W(|\overline{\boldsymbol{x}_i} - \overline{\boldsymbol{x}_j}|, \sqrt{2}h)\frac{m}{\overline{\rho}}.
\end{aligned} \tag{2.135}$$

DR for the cubic spline interpolation is expressed as

$$\omega^2_{\text{cubic}} = -C_s^2 Da + \frac{\overline{P}}{\overline{\rho}}\left[2Da + 2b - \frac{1}{2}h^2 C_\rho c + \frac{1}{2}C_\rho d\right], \tag{2.136}$$

where

$$c = \sum_{j \neq i} \frac{\overline{x_i} - \overline{x_j}}{|\overline{\boldsymbol{x}_i} - \overline{\boldsymbol{x}_j}|^2}(1 - \cos[k(\overline{x_i} - \overline{x_j})])\frac{\partial}{\partial \overline{x_i}} W(|\overline{\boldsymbol{x}_i} - \overline{\boldsymbol{x}_j}|, \sqrt{2}h)\frac{m}{\overline{\rho}},$$

$$d = \sum_{j \neq i} (\overline{x_i} - \overline{x_j})(1 - \cos[k(\overline{x_i} - \overline{x_j})])\frac{\partial}{\partial \overline{x_i}} W(|\overline{\boldsymbol{x}_i} - \overline{\boldsymbol{x}_j}|, \sqrt{2}h)\frac{m}{\overline{\rho}}. \tag{2.137}$$

DR for the quintic spline interpolation is expressed as

$$\begin{aligned}\omega^2_{\text{quintic}} =& - C_s^2 Da + \frac{\overline{P}}{\overline{\rho}}\left[2Da + 2b + \frac{3}{8}h^4 C_\rho B_{1,4} + \frac{3}{16}h^4 C_\rho D A_{2,4} - \frac{3}{4}h^2 C_\rho B_{1,2}\right.\\ &\left. - \frac{1}{8}h^2 C_\rho D A_{2,2} + \frac{5}{8}C_\rho B_{1,0} + \frac{1}{16}C_\rho D A_{2,0}\right],\end{aligned} \tag{2.138}$$

where

$$A_{n,m} = \sum_j \frac{(\overline{x_i} - \overline{x_j})^n}{|\overline{\boldsymbol{x}_i} - \overline{\boldsymbol{x}_j}|^m} \sin[k(\overline{x_i} - \overline{x_j})]\frac{\partial}{\partial \overline{x_i}} W(|\overline{\boldsymbol{x}_i} - \overline{\boldsymbol{x}_j}|, \sqrt{2}h)\frac{m}{\overline{\rho}},$$

$$B_{n,m} = \sum_j \frac{(\overline{x_i} - \overline{x_j})^n}{|\overline{\boldsymbol{x}_i} - \overline{\boldsymbol{x}_j}|^m}(1 - \cos[k(\overline{x_i} - \overline{x_j})])\frac{\partial}{\partial \overline{x_i}} W(|\overline{\boldsymbol{x}_i} - \overline{\boldsymbol{x}_j}|, \sqrt{2}h)\frac{m}{\overline{\rho}}. \tag{2.139}$$

References

J. Barnes, P. Hut, A hierarchical O(N log N) force-calculation algorithm. Nature **324**, 446–449 (1986). https://doi.org/10.1038/324446a0

W. Benz, E. Asphaug, Impact simulations with fracture. I. Method and tests. Icarus **107**, 98 (1994). https://doi.org/10.1006/icar.1994.1009

W. Benz, E. Asphaug, Simulations of brittle solids using smooth particle hydrodynamics. Comput. Phys. Commun. **87**, 253–265 (1995). https://doi.org/10.1016/0010-4655(94)00176-3

W. Benz, E. Asphaug, Catastrophic disruptions revisited. Icarus **142**, 5–20 (1999). https://doi.org/10.1006/icar.1999.6204

J. Bonet, T.-S. Lok, Variational and momentum preservation aspects of smooth particle hydrodynamic formulations. Comput. Methods Appl. Mech. Eng. **180**(1), 97–115 (1999). https://doi.org/10.1016/S0045-7825(99)00051-1

J.K. Chen, J.E. Beraun, C.J. Jih, An improvement for tensile instability in smoothed particle hydrodynamics. Comput. Mech. **23**, 279–287 (1999). https://doi.org/10.1007/s004660050409

G.S. Collins, H.J. Melosh, B.A. Ivanov, Modeling damage and deformation in impact simulations. Meteorit. Planet. Sci. **39**, 217–231 (2004). https://doi.org/10.1111/j.1945-5100.2004.tb00337.x

H. Genda, E. Kokubo, S. Ida, Merging criteria for giant impacts of protoplanets. Astrophys. J. **744**, 137 (2012). https://doi.org/10.1088/0004-637X/744/2/137

R.A. Gingold, J.J. Monaghan, Smoothed particle hydrodynamics: theory and application to non-spherical stars. Mon. Not. R. Astron. Soc. **181**, 375–389 (1977). https://doi.org/10.1093/mnras/181.3.375

D. Grady, M. Kipp, Continuum modelling of explosive fracture in oil shale. Int. J. Rock Mech. Min. Sci. Geomech. Abstr. **17**(3), 147–157 (1980). https://doi.org/10.1016/0148-9062(80)91361-3

J.P. Gray, J.J. Monaghan, R.P. Swift, Sph elastic dynamics. Comput. Methods Appl. Mech. Eng. **190**(49), 6641–6662 (2001). https://doi.org/10.1016/S0045-7825(01)00254-7

S. Inutsuka, Reformulation of smoothed particle hydrodynamics with Riemann solver. J. Comput. Phys. **179**, 238–267 (2002). https://doi.org/10.1006/jcph.2002.7053

K. Iwasaki, S.-I. Inutsuka, Smoothed particle magnetohydrodynamics with a riemann solver and the method of characteristics. Mon. Not. R. Astron. Soc. **418**(3), 1668–1688 (2011). https://doi.org/10.1111/j.1365-2966.2011.19588.x

M. Iwasawa, A. Tanikawa, N. Hosono, K. Nitadori, T. Muranushi, J. Makino, Fdps: a novel framework for developing high-performance particle simulation codes for distributed-memory systems, in *Proceedings of the 5th International Workshop on Domain-Specific Languages and High-Level Frameworks for High Performance Computing*, WOLFHPC '15, (ACM, New York, NY, USA, 2015), pp. 1:1–1:10 ISBN 978-1-4503-4016-8. https://doi.org/10.1145/2830018.2830019

M. Iwasawa, A. Tanikawa, N. Hosono, K. Nitadori, T. Muranushi, J. Makino, Implementation and performance of FDPS: a framework for developing parallel particle simulation codes. Publ. Astron. Soc. Japan **68**(4), 24 (2016)

M. Jutzi, SPH calculations of asteroid disruptions: the role of pressure dependent failure models. Planet. Space Sci. **107**, 3–9 (2015). https://doi.org/10.1016/j.pss.2014.09.012

M. Jutzi, E. Asphaug, The shape and structure of cometary nuclei as a result of low-velocity accretion. Science **348**, 1355–1358 (2015). https://doi.org/10.1126/science.aaa4747

E. Lajeunesse, J.B. Monnier, G.M. Homsy, Granular slumping on a horizontal surface. Phys. Fluids **17**(10), 103302 (2005). https://doi.org/10.1063/1.2087687

M.A. Lange, T.J. Ahrens, The dynamic tensile strength of ice and ice-silicate mixtures. J. Geophys. Res. **88**(B2), 1197–1208 (1983). https://doi.org/10.1029/JB088iB02p01197

B. Lawn. *Fracture of Brittle Solids*. Cambridge Solid State Science Series, 2nd edn. (Cambridge University Press, 1993). https://doi.org/10.1017/CBO9780511623127

A. Leleu, M. Jutzi, M. Rubin, The peculiar shapes of Saturn's small inner moons as evidence of mergers of similar-sized moonlets. Nat. Astron. **2**, 555–561 (2018). https://doi.org/10.1038/s41550-018-0471-7

L.D. Libersky, A.G. Petschek, Smooth particle hydrodynamics with strength of materials, in *Advances in the Free-Lagrange Method Including Contributions on Adaptive Gridding and the Smooth Particle Hydrodynamics Method*, ed. H.E. Trease, M.F. Fritts, W.P. Crowley, vol. 395 of *Lecture Notes in Physics* (Berlin Springer Verlag, 1991), pp. 248–257

E.M. Lifshitz, A.M. Kosevich, L.P. Pitaevskii, Chapter ii : the equilibrium of rods and plates. in *Theory of Elasticity*, 3rd edn, ed. by E. Lifshitz, A. Kosevich, L. Pitaevskii (Butterworth-Heinemann, Oxford, 1986), pp. 38–86. ISBN 978-0-08-057069-3. https://doi.org/10.1016/B978-0-08-057069-3.50009-7

L.B. Lucy, A numerical approach to the testing of the fission hypothesis. Astron. J. **82**, 1013–1024 (1977). https://doi.org/10.1086/112164

H.J. Melosh, E.V. Ryan, E. Asphaug, Dynamic fragmentation in impacts: hydrocode simulation of laboratory impacts. J. Geophys. Res. **97**, 14 (1992). https://doi.org/10.1029/92JE01632

J.J. Monaghan, An introduction to SPH. Comput. Phys. Commun. **48**, 89–96 (1988). https://doi.org/10.1016/0010-4655(88)90026-4

J.J. Monaghan, Smoothed particle hydrodynamics. Annu. Rev. Astron. Astrophys. **30**, 543–574 (1992). https://doi.org/10.1146/annurev.aa.30.090192.002551

J.J. Monaghan, SPH without a Tensile Instability. J. Comput. Phys. **159**, 290–311 (2000). https://doi.org/10.1006/jcph.2000.6439

A. Nakamura, A. Fujiwara, Velocity Distribution of Fragments in Collisional Breakup. in *IAU Colloq. 126: Origin and Evolution of Interplanetary Dust*, vol. 173 of *Astrophysics and Space Science Library*, ed. by A.C. Levasseur-Regoud, H. Hasegawa (1991), p. 379

J.D. O'Keefe, T.J. Ahrens, Cometary and meteorite swarm impact on planetary surfaces. J. Geophys. Res. **87**, 6668–6680 (1982a). https://doi.org/10.1029/JB087iB08p06668

J.D. O'Keefe, T.J. Ahrens, *The interaction of the Cretaceous/Tertiary Extinction Bolide with the atmosphere, ocean, and solid Earth* (1982b). https://doi.org/10.1130/SPE190

P.W. Randles, L.D. Libersky, Smoothed particle hydrodynamics: some recent improvements and applications. Comput. Methods Appl. Mech. Eng. **139**(1), 375–408 (1996). https://doi.org/10.1016/S0045-7825(96)01090-0

L.D.G. Sigalotti, H. López, Adaptive Kernel estimation and SPH tensile instability. Comput. Math. Appl **55**(1), 23–50 (2008). https://doi.org/10.1016/j.camwa.2007.03.007

K. Sugiura, S. Inutsuka, An extension of Godunov SPH: application to negative pressure media. J. Comput. Phys. **308**, 171–197 (2016). https://doi.org/10.1016/j.jcp.2015.12.030

K. Sugiura, S. Inutsuka, An extension of Godunov SPH II: application to elastic dynamics. J. Comput. Phys. **333**, 78–103 (2017). https://doi.org/10.1016/j.jcp.2016.12.026

K. Sugiura, H. Kobayashi, S. Inutsuka, Toward understanding the origin of asteroid geometries. Variety in shapes produced by equal-mass impacts. Astron. Astrophys. **620**, A167 (2018). https://doi.org/10.1051/0004-6361/201833227

J.W. Swegle, D.L. Hicks, S.W. Attaway, Smoothed particle hydrodynamics stability analysis. J. Comput. Phys. **116**, 123–134 (1995). https://doi.org/10.1006/jcph.1995.1010

J.H. Tillotson, Metallic equations of state for hypervelocity impact. General At. Rept. **3216**, 1–142 (1962)

B.V. Leer, Towards the ultimate conservative difference scheme. v. a second-order sequel to Godunov's method. J. Comput. Phys. **32**(1), 101–136 (1979). https://doi.org/10.1016/0021-9991(79)90145-1

J.B. Walsh, The effect of cracks on the compressibility of rocks. J. Geophys. Res. **70**, 381–389 (1965). https://doi.org/10.1029/JZ070i002p00381

W.A. Weibull, A statistical theory of the strength of materials (transl.). Ingvetenks. Akad. Handl. (Stockholm) **151**, 5–45 (1939)

Chapter 3
Results: Shapes of Impact Outcomes

Abstract In this chapter, we introduce the results of our simulations that investigate shapes of asteroids formed through various collisions between asteroids. Database of Asteroid Models from Inversion Techniques provides almost all shapes of known asteroids with diameters larger than 100 km. Thus, in our impact simulations, we set radius of target asteroids to 50 km to allow us to directly compare the results of our simulations with the shapes of actual asteroids. We vary impact angles, impact velocities, and ratios of the mass of impactor asteroids to that of target asteroids, and we conduct various impact simulations. As a result, we find that similar-mass and low-velocity impacts produce various shapes including extremely flat and elongated shapes. In contrast, high-velocity and destructive impacts mainly produce spherical and bilobed shapes, but they are difficult to produce flat shapes.

Keywords Numerical simulations · Collisions between asteroids · Shapes of collisional outcomes

3.1 Initial Conditions of Impacts

We use basaltic spheres as asteroids before collisions. We ignore initial rotations of the initial asteroids. Radius of target asteroids is fixed to $R_t = 50$ km, and thus mass of target asteroids is $M_{\text{target}} = (4/3)\pi\rho_0 R_t^3$, where ρ_0 is uncompressed density of basaltic materials and is the same as ρ_0 of the Tillotson parameter.

For asteroids with radius $\leq$50 km in hydrostatic equilibrium, even the central density is very similar to ρ_0 because of insignificant compression due to the weak self-gravity. Thus, we use uniform spheres with ρ_0 as initial asteroids. We prefer uniform SPH particle distribution but with small random dispositions to eliminate influences caused by systematic particle distribution. We prepare such uniform particle distribution from highly random particle distribution. For detailed procedures to produce the uniform distribution, see Sugiura et al. (2018).

We define the impact velocity v_{imp} and the impact angle θ_{imp} as shown in Fig. 3.1. Thus, $\theta_{\text{imp}} = 0°$ means a head-on collision, and $\theta_{\text{imp}} = 90°$ means a complete grazing collision. We also define the mass ratio $q = M_i/M_{\text{target}}$, where $M_i = (4/3)\pi R_i^3 \rho_0$ is

K. Sugiura, *Development of a Numerical Simulation Method for Rocky Body Impacts and Theoretical Analysis of Asteroidal Shapes*, Springer Theses,
https://doi.org/10.1007/978-981-15-3722-6_3

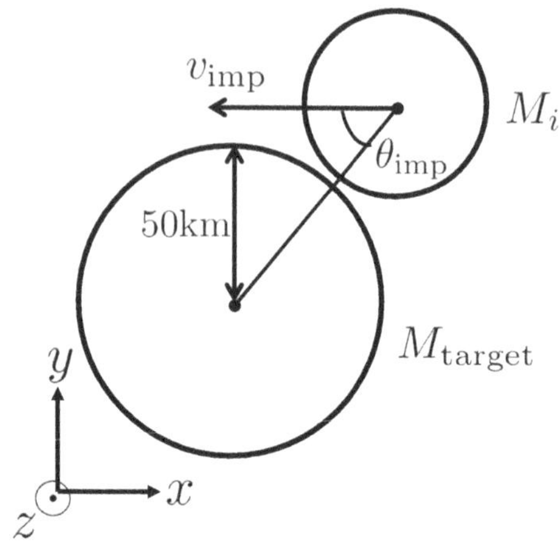

Fig. 3.1 Relation between impact geometries and coordinate system used in our simulations. This figure also shows the definition of the impact velocity v_{imp} and the impact angle θ_{imp}

mass of impactor asteroids and R_i is radius of impactor asteroids. Initial distances between centers of two asteroids are set to $2(R_t + R_i)$.

We assume that friction angle for completely damaged material is $\phi_d = 40°$, which corresponds to the friction coefficient of $\mu_d = \tan(\phi_d) = 0.839$. Note that the measured friction angle for lunar soil is $30°–50°$ (e.g., Heiken et al. 1991). $\phi_d = 40°$ is also consistent with the maximum surface slopes of the asteroid Itokawa $\sim 40°$ (Fujiwara et al. 2006).

3.2 Analysis of Results

Each impact simulation is continued over a period of 1.0×10^5 s, which is much longer than the timescale of the gravitational reaccumulation. The reaccumulation timescale t_{acc} is estimated as

$$t_{acc} = \frac{2R_t}{v_{esc,t}} = \sqrt{\frac{3}{2\pi G \rho_0}} \simeq 1600\,\text{s}, \tag{3.1}$$

where $v_{esc,t}$ is the escape velocity from target asteroids. Thus the computational period is about 100 times longer than t_{acc}, so that the gravitational reaccumulation of fragments produced by impacts is sufficiently finished at the end of simulations.

To evaluate shapes of remnants produced through impacts, we firstly identify remnants made up of SPH particles using a friends-of-friends algorithm (e.g., Huchra and Geller 1982). Swarms of SPH particles with spacing less than $1.5h$ are identified with remnants.

Then we measure axis lengths of remnants using the inertia moment. The inertia moment tensor of a remnant is calculated as

$$I^{\alpha\beta} = \sum_k m_k \left[(x_k^\gamma - x_{\rm CoM}^\gamma)(x_k^\gamma - x_{\rm CoM}^\gamma)\delta^{\alpha\beta} - (x_k^\alpha - x_{\rm CoM}^\alpha)(x_k^\beta - x_{\rm CoM}^\beta) \right], \quad (3.2)$$

where $x_{\rm CoM}$ shows the center of mass of the remnant and k shows SPH particles that compose the remnant. The principal moments of inertia $I_1 > I_2 > I_3$ are obtained from $I^{\alpha\beta}$. Here, the remnant is approximated as an uniform ellipsoid with the same inertia moment $I^{\alpha\beta}$ and the mass M. The three principal moments of inertia of the ellipsoid are expressed as

$$\begin{aligned} I_1 &= \frac{1}{20}(a^2 + b^2)M, \\ I_2 &= \frac{1}{20}(a^2 + c^2)M, \\ I_3 &= \frac{1}{20}(b^2 + c^2)M, \end{aligned} \quad (3.3)$$

where a, b, and c represent the major axis length, the intermediate axis length, and the minor axis length, respectively. Equation (3.3) is rewritten as

$$\begin{aligned} a &= \sqrt{\frac{10(I_1 + I_2 - I_3)}{M}}, \\ b &= \sqrt{\frac{10(I_1 - I_2 + I_3)}{M}}, \\ c &= \sqrt{\frac{10(-I_1 + I_2 + I_3)}{M}}. \end{aligned} \quad (3.4)$$

Finally, we identify the axis lengths a, b, and c of the ellipsoid with those of the remnant. Two axis ratios b/a and c/a specify shapes, i.e., spherical shapes have $b/a \approx 1$ and $c/a \approx 1$, flat shapes have $b/a \approx 1$ and $c/a \ll 1$, and elongated shapes have $b/a \ll 1$ and $c/a \ll 1$.

We have other measurement methods of the axis ratios such as the bottom-up method, which is usually used in laboratory experiments (e.g., Fujiwara et al. 1978). In this method, the axis lengths of a remnant are defined as distances between a pair of parallel plates that contact with the remnant. Detailed procedures to measure the axis lengths in the bottom-up method are as follows: Firstly, we identify the minor axis length c with the minimum distance among distances between parallel plates, and define the direction normal to the plates when the distance is minimum as the minor axis. Secondly, we identify the intermediate axis length b with the minimum distance among distances between parallel plates when the directions normal to the plates are perpendicular to the minor axis, and also define the intermediate axis. Finally, we identify the major axis length a as the distance between the parallel plates when the direction normal to the plates is perpendicular to both the minor and intermediate axes.

Note that the axis ratios obtained from different measurement methods are slightly different since bodies are, of course, not perfect ellipsoids. As shown in Appendix A of Michikami et al. (2018), the difference of axis ratios obtained from different methods is at most about 0.1; the axis ratios include measurement errors of ~0.1. Thus, if difference of the axis ratios of two different remnants is larger than 0.1, these remnants have quantitatively different shapes.

3.3 Equal-Mass and Non-destructive Impacts

In this section, we introduce results of our simulations of equal-mass and non-destructive impacts. We set the mass of impactor asteroids to $M_i = M_{\rm target}$, and we conduct various impact simulations with various impact velocities and angles. We focus on non-destructive impacts that have mass of the largest remnants larger than $0.4M_{\rm target}$, and investigate shapes of the largest remnants. The total number of SPH particles for each impact simulation is set to 100,000, which is relatively low resolution but enables to investigate shapes of remnants produced through various impacts with various impact velocities and angles. The resolution dependence of resultant shapes is investigated in Sect. 3.3.1, and we show that this number of SPH particles is sufficient to investigate shapes of, at least, the largest remnants.

3.3.1 Resolution Dependence on Resultant Shapes

Figure 3.2 shows the result of the simulation with the impact angle $\theta_{\rm imp}$ of 15° and the impact velocity $v_{\rm imp}$ of 200 m/s. The collision shatters the asteroids (Fig. 3.2b) and stretches the asteroids into the elongated configuration (Fig. 3.2c). The self-gravity prevents further elongation and ejected materials reaccumulate on the long axis of the largest remnant (Fig. 3.2d), which finally results in the formation of a very elongated shape with the ratio $b/a \approx 0.2$ (Fig. 3.2f).

Figure 3.3 shows the shapes of the largest remnants produced through the impacts with the same condition but with three different resolutions. The shapes of the largest remnants with $N_{\rm total} = 5 \times 10^4, 2 \times 10^5$, and 8×10^5 are almost the same. Figure 3.4 shows the mass and the axis ratios of the largest remnants produced through these three simulations. The mass of the largest remnant is almost constant for these simulations. Although the axis ratios are slightly larger for larger $N_{\rm total}$, the difference of b/a between $N_{\rm total} = 5 \times 10^4$ and 8×10^5 is about 0.03, which is a minor difference for the analysis of asteroidal shapes (see Sect. 3.2). Thus, $N_{\rm total} = 10^5$ is sufficient to capture the feature of resultant shapes produced through non-destructive impacts.

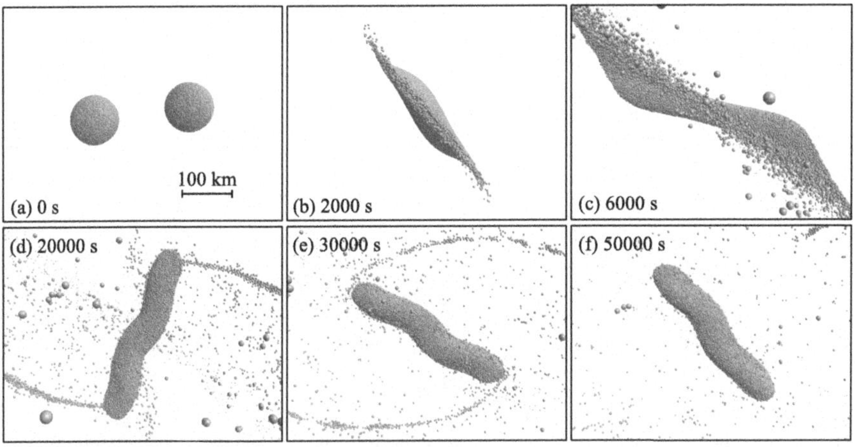

Fig. 3.2 Snapshots of the impact simulation with the impact angle $\theta_{\mathrm{imp}} = 15°$, the impact velocity $v_{\mathrm{imp}} = 200\,\mathrm{m/s}$ and the total number of SPH particles $N_{\mathrm{total}} = 1 \times 10^5$. All the panels have the same scale shown in the panel (**a**). Sugiura et al., A&A, 620, A167, 2018, reproduced with permission ©ESO

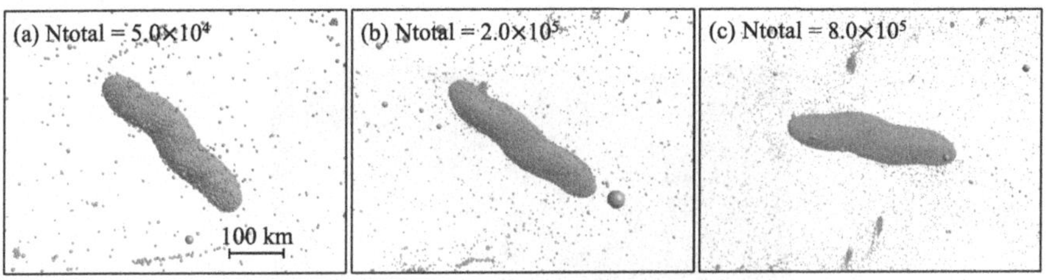

Fig. 3.3 Snapshots of the impacts with the same condition of $v_{\mathrm{imp}} = 200\,\mathrm{m/s}$ and $\theta_{\mathrm{imp}} = 15°$ but with the different resolutions of $N_{\mathrm{total}} = 5 \times 10^4$ (**a**), 2×10^5 (**b**), and 8×10^5 (**c**), respectively. These snapshots are taken at 5.0×10^4 s. Sugiura et al., A&A, 620, A167, 2018, reproduced with permission ©ESO

3.3.2 Mass of the Largest Remnants

In the rest of Sect. 3.3, we set $N_{\mathrm{total}} = 10^5$ and measure the mass and the axis ratios of the largest remnants at 1.0×10^5 s.

Figure 3.5 summarizes the mass of the largest remnants M_{lr} produced through various impacts. We vary v_{imp} with the range of 50–400 m/s and the increment of 25 m/s, and we vary θ_{imp} with the range of $5°$–$45°$ and the increment of $5°$. For $\theta_{\mathrm{imp}} \leq 15°$, impacts with low v_{imp} result in collisional merging ($M_{\mathrm{lr}} \sim 2M_{\mathrm{target}}$), and those with higher v_{imp} result in erosive collisions and M_{lr} gradually decreases with increasing v_{imp}. In contrast, for $\theta_{\mathrm{imp}} \geq 20°$, impacts with high v_{imp} result in "hit-and-run" processes, which mainly have $M_{\mathrm{lr}} \sim M_{\mathrm{target}}$, and thus we observe sharp variation in M_{lr} around $v_{\mathrm{imp}} \sim 100\,\mathrm{m/s}$. Previous collisional simulations also show

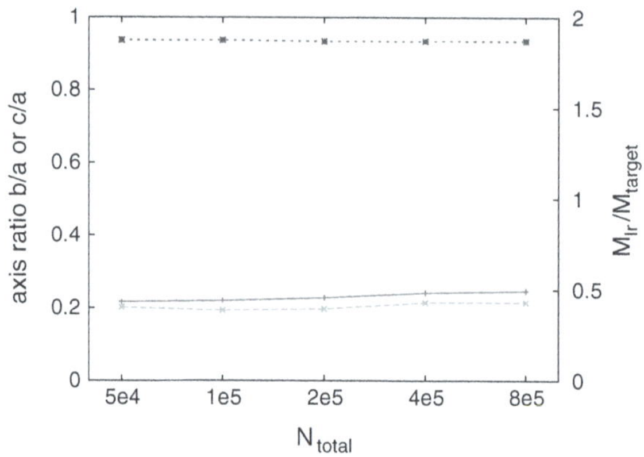

Fig. 3.4 Mass and axis ratios of the largest remnants produced through the impacts with the same condition of $\theta_{imp} = 15°$ and $v_{imp} = 200\,m/s$ but with the different resolutions N_{total}. The red solid line, the green dashed line, and the blue dotted line show b/a, c/a (refer to the left vertical axis), and the mass of the largest remnants M_{lr} normalized by M_{target} (refer to the right vertical axis), respectively. Sugiura et al., A&A, 620, A167, 2018, reproduced with permission ©ESO

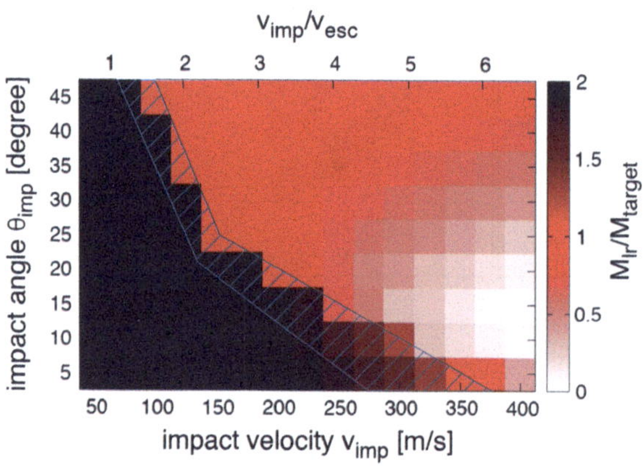

Fig. 3.5 Mass of the largest remnants M_{lr} produced through impact simulations with various v_{imp} (horizontal axis) and θ_{imp} (vertical axis). The upper horizontal axis represents v_{imp} normalized by the two-body escape velocity v_{esc}. The color shows M_{lr}/M_{target}, and thus $M_{lr}/M_{target} = 2.0$ (black color) means complete merging. Approximate transitional parameter range from merging collisions to hit-and-run or elosive collisions is indicated by the hatched region. Sugiura et al., A&A, 620, A167, 2018, reproduced with permission ©ESO

that erosive nature for low impact angles and merging/hit-and-run nature for high impact angles (Agnor and Asphaug 2004; Leinhardt and Stewart 2012).

It is instructive to note that, for $v_{\rm imp} > 300\,{\rm m/s}$, $M_{\rm lr}$ shows the lowest value not at $\theta_{\rm imp} = 0°$ but at $\theta_{\rm imp} \approx 15°$. Head-on collisions ($\theta_{\rm imp} = 0°$) yield large $M_{\rm lr}$ because the most of the impact energy is dissipated, converted to internal energy, and not used for the ejection of fragments. In contrast, impacts with slightly higher $\theta_{\rm imp}$ result in more efficient ejection of fragments because of larger shear component of impact velocities. However, for much higher $\theta_{\rm imp}$, small portions of bodies directly collide and only material in these portions are ejected, which results in larger $M_{\rm lr}$. Thus, the intermediate $\theta_{\rm imp} \approx 15°$ has the minimum $M_{\rm lr}$.

For $\theta_{\rm imp} > 45°$, only edges of bodies are destructed and their shapes do not deform so much. Thus we do not investigate resultant shapes produced through impacts with such high impact angles. Impacts with $v_{\rm imp} > 400\,{\rm m/s}$ mainly result in catastrophic disruption, and the resolution of $N_{\rm total} = 1 \times 10^5$ is not sufficient to investigate shapes of remnants resulted from such destructive impacts. Results of high-resolution simulations of catastrophic disruption are mainly shown in Sect. 3.5.

3.3.3 *Characteristic Shapes Formed Through Collisions*

We find that resultant shapes of the largest remnants produced through the impact simulations introduced in Sect. 3.3.2 ($50\,{\rm m/s} \le v_{\rm imp} \le 400\,{\rm m/s}$ and $5° \le \theta_{\rm imp} \le 45°$) are roughly classified into five categories. Here, we present snapshots of impact simulations that produce shapes classified into five different categories, a catastrophic collision, and an impact with the second collision.

3.3.3.1 Bilobed Shapes

Figure 3.6 shows the result of the impact simulation with $v_{\rm imp} = 50\,{\rm m/s}$ and $\theta_{\rm imp} = 30°$. The spherical shapes of the initial colliders do not deform with the very low impact velocity $v_{\rm imp} = 50\,{\rm m/s}$ and they just stick together (Fig. 3.6b). Such gentle collision produces a bilobed shape (Fig. 3.6d). The impact velocity of this simulation is slower than the two-body escape velocity $v_{\rm esc} \approx 60\,{\rm m/s}$, and the colliders are initially bounded by the mutual self gravity. Thus the formation of bilobed shapes through such low velocity impacts does not depend on the impact angle $\theta_{\rm imp}$.

3.3.3.2 Spherical Shapes

Figure 3.7 shows the result of the impact simulation with $v_{\rm imp} = 100\,{\rm m/s}$ and $\theta_{\rm imp} = 10°$, which produces a spherical shape. The impact with slightly higher $v_{\rm imp} = 100\,{\rm m/s}$ deforms the initial spherical shapes but still results in the collisional merging (Fig. 3.7b). The minor deformation produces the spherical shape (Fig. 3.7c, d).

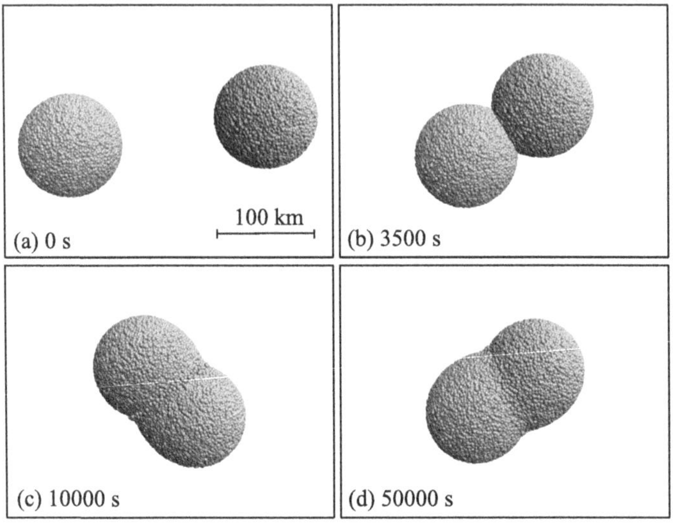

Fig. 3.6 Snapshots of the impact simulation with $\theta_{imp} = 30°$ and $v_{imp} = 50\,m/s$. Sugiura et al., A&A, 620, A167, 2018, reproduced with permission ©ESO

Fig. 3.7 Snapshots of the impact simulation with $\theta_{imp} = 10°$ and $v_{imp} = 100\,m/s$. Sugiura et al., A&A, 620, A167, 2018, reproduced with permission ©ESO

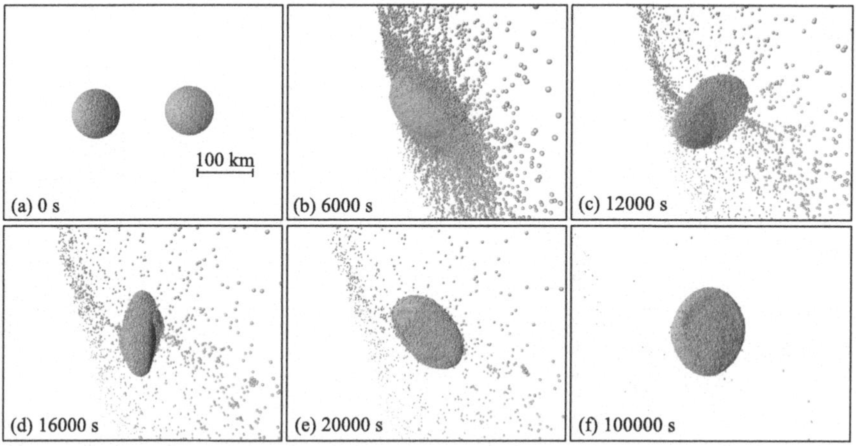

Fig. 3.8 Snapshots of the impact simulation with $\theta_{\rm imp} = 5°$ and $v_{\rm imp} = 200\,{\rm m/s}$. Sugiura et al., A&A, 620, A167, 2018, reproduced with permission ©ESO

Note that impacts with $v_{\rm imp} \approx 100\,{\rm m/s}$ and $\theta_{\rm imp} \geq 40°$ result in hit-and-run collisions with local excavation. Resultant shapes of these impacts are also close to spherical shapes.

3.3.3.3 Flat Shapes

Figure 3.8 shows the snapshots of the impact with $v_{\rm imp} = 200\,{\rm m/s}$ and $\theta_{\rm imp} = 5°$. The impact with the higher impact velocity $v_{\rm imp} = 200\,{\rm m/s}$ results in more major deformation (Fig. 3.8b, c), and a flat shape is formed (Fig. 3.8d–f). The ratio of the minor to major axis lengths c/a of the flat shape is about 0.4. The direction of the minor axis is perpendicular to the rotation axis.

3.3.3.4 Elongated Shapes

Figure 3.2 in Sect. 3.3.1 shows the merging collision that produces the elongated shape. Elongated shapes are also produced through some hit-and-run collisions. Figure 3.9 shows the result of the impact simulation with $v_{\rm imp} = 250\,{\rm m/s}$ and $\theta_{\rm imp} = 20°$, and Fig. 3.10 shows the zoom out view of Fig. 3.9f. The impact stretches the bodies into a filament (Fig. 3.9c), and the higher impact velocity compared to that in Fig. 3.2 results the fragmentation of the filament (Fig. 3.9c). Two elongated shapes are formed through the reaccumulation of fragments (Figs. 3.9d–f and 3.10). Note that the largest and the second largest objects produced through equal-mass and hit-and-run collisions have very similar shapes because of symmetric impact configurations (Fig. 3.10).

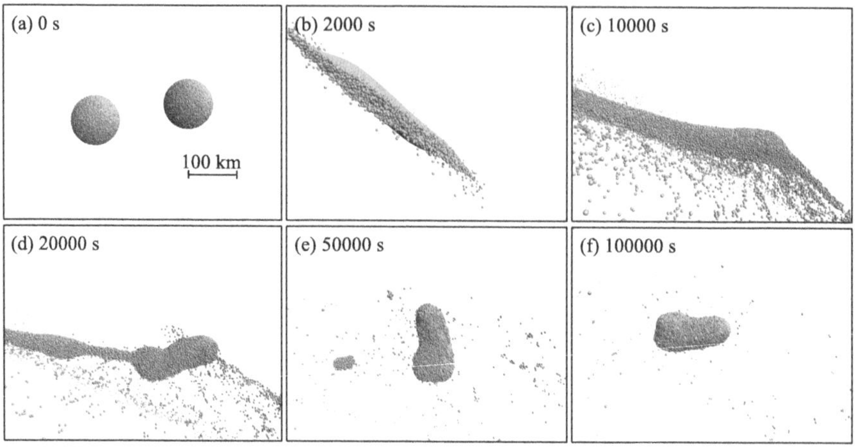

Fig. 3.9 Snapshots of the impact simulation with $\theta_{\rm imp} = 20°$ and $v_{\rm imp} = 250\,{\rm m/s}$. Sugiura et al., A&A, 620, A167, 2018, reproduced with permission ©ESO

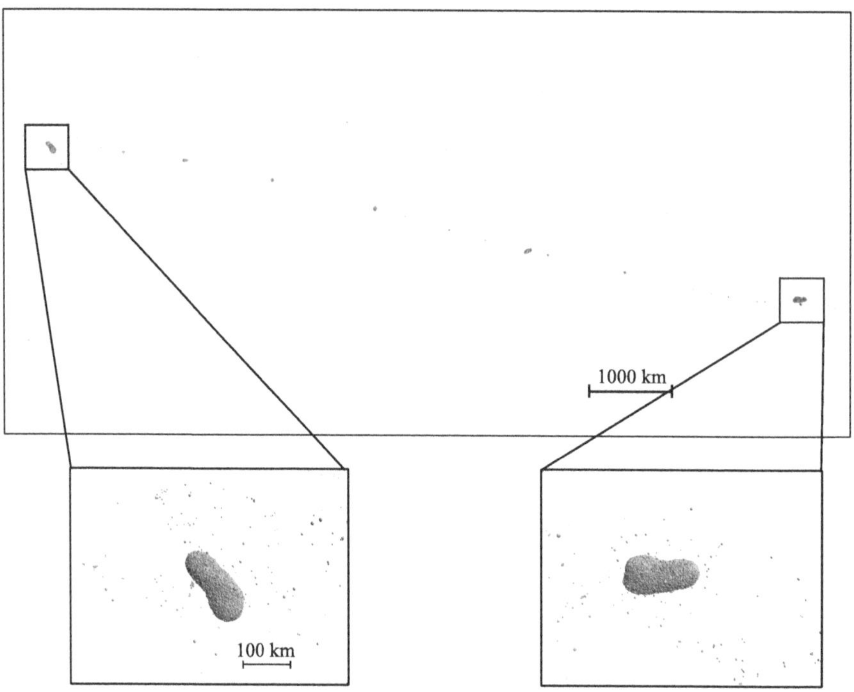

Fig. 3.10 Zoom out view of the impact simulation with $v_{\rm imp} = 250\,{\rm m/s}$ and $\theta_{\rm imp} = 20°$ at 1.0×10^5 s. Two enlarged snapshots show the largest and the second largest remnants. Sugiura et al., A&A, 620, A167, 2018, reproduced with permission ©ESO

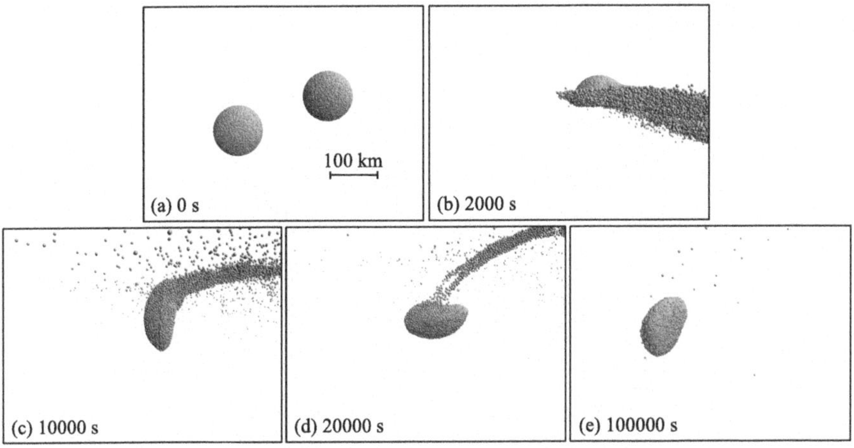

Fig. 3.11 Snapshots of the impact simulation with $\theta_{\mathrm{imp}} = 45°$ and $v_{\mathrm{imp}} = 350\,\mathrm{m/s}$. Sugiura et al., A&A, 620, A167, 2018, reproduced with permission ©ESO

3.3.3.5 Hemispherical Shapes

Figure 3.11 shows the impact with the higher impact angle $\theta_{\mathrm{imp}} = 45°$ that forms hemispherical shapes. Significant amount of fragments are ejected from the impact points, but the opposite sides of the impact points are unaffected (Fig. 3.11b). The excavation of the halves of the bodies produces the hemispherical shapes (Fig. 3.11c–e).

3.3.3.6 Super-Catastrophic Destruction

Figure 3.12 represents the snapshots of the impact with $v_{\mathrm{imp}} = 400\,\mathrm{m/s}$ and $\theta_{\mathrm{imp}} = 5°$. The large ejecta curtain is produced through the impact with the very high impact velocity (Fig. 3.12b), and many clumps are formed through the gravitational reaccumulation of fragments on the curtain (Fig. 3.12c). Then the gravitational reaccumulation of many clumps and fragments produces the largest remnant (Fig. 3.12d–f).

The formation process dominated by the coalescence of many small clumps is remarkable for impacts with $M_{\mathrm{lr}} < 0.4 M_{\mathrm{target}}$. Here, we call such impacts super-catastrophic destruction. A super-catastrophic destruction produces many small remnants and the largest remnant is not special one; we need to statistically investigate the distribution of shapes of many remnants through high-resolution simulations. Thus, we do not investigate shapes of the largest remnants for super-catastrophic destruction here, and leave this to Sect. 3.5.

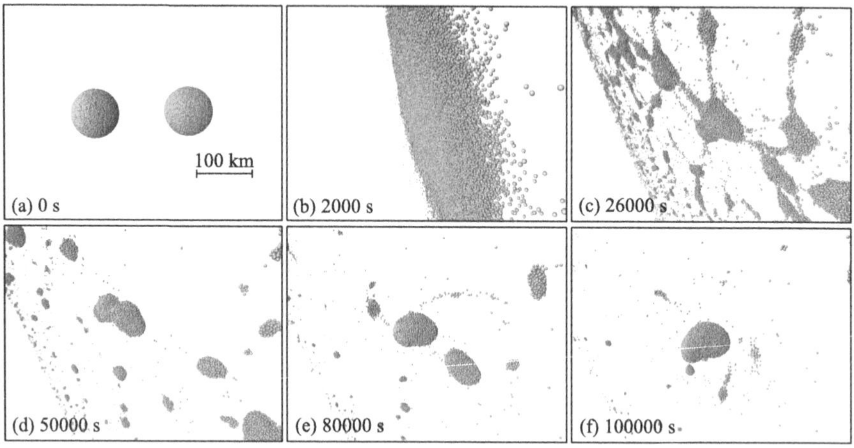

Fig. 3.12 Snapshots of the impact simulation with $v_{\rm imp} = 400\,{\rm m/s}$ and $\theta_{\rm imp} = 5°$. Sugiura et al., A&A, 620, A167, 2018, reproduced with permission ©ESO

3.3.3.7 Elongated-Shape-Forming Impacts with the Second Collision

Figure 3.13 shows the snapshots of the collision with $v_{\rm imp} = 175$ m/s and $\theta_{\rm imp} = 20°$. The two elongated objects are formed by the first collision like the impact shown in Fig. 3.9 (Fig. 3.13b, c). Although the impact velocity is larger than the two-body escape velocity, the collision dissipates the kinetic energy and then two colliding bodies are gravitationally bounded (Fig. 3.13d). The final body formed by the second collision is no longer elongated object (Fig. 3.13e), but this impact is a kind of elongated-shape-forming collision as we observe the elongated shapes after the first collision.

3.3.4 Summary of Shapes Formed by Collisions

Figure 3.14 shows the axis ratios of the largest remnants produced through various impacts. Hit-and-run collisions produce the largest and the second largest remnants with similar masses and shapes (see Fig. 3.10) and typically result in the ratio of the largest mass to the second largest mass smaller than 2.0. Thus we show the averaged b/a and c/a among the largest and the second largest bodies for such hit-and-run collisions.

Figure 3.15 shows the summary of the axis ratios of the largest remnants produced through the non-destructive impacts with $50\,{\rm m/s} \leq v_{\rm imp} \leq 400\,{\rm m/s}$, $5° \leq \theta_{\rm imp} \leq 45°$, and $M_{\rm lr} > 0.4M_{\rm target}$. From Fig. 3.15, we notice that the equal-mass and non-destructive impacts produce extremely elongated shapes with $b/a < 0.4$ and extremely flat shapes with $b/a > 0.8$ and $c/a < 0.6$.

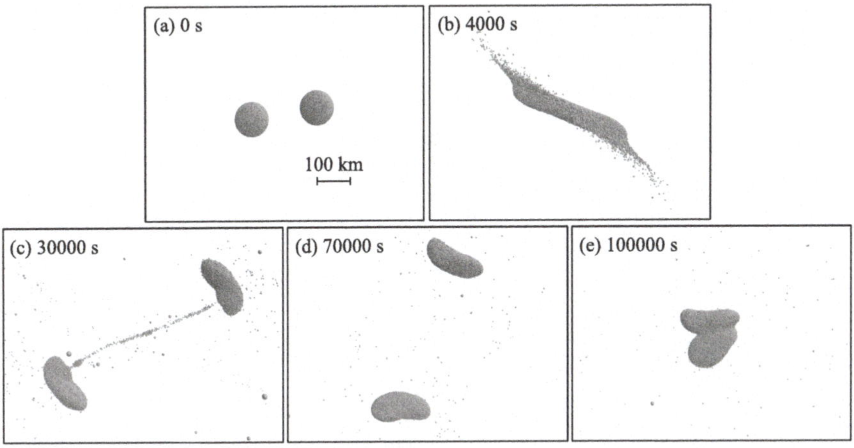

Fig. 3.13 Snapshots of the impact simulation with $v_{\text{imp}} = 175\,\text{m/s}$ and $\theta_{\text{imp}} = 20°$. Sugiura et al., A&A, 620, A167, 2018, reproduced with permission ©ESO

Fig. 3.14 Axis ratios b/a (**a**) and c/a (**b**) of the largest remnants produced through various impacts with various v_{imp} (horizontal axis) and θ_{imp} (vertical axis). The cross-hatched region represents catastrophic collisions with $M_{\text{lr}} < 0.15 M_{\text{target}}$, and the axis ratios of the largest remnants are not measured for these impacts. The hatched regions again show the transitions from merging collisions to hit-and-run/elosive collisions (see Fig. 3.5). Two green boxes show impacts accompanied by the second collision as shown in Fig. 3.13. Sugiura et al., A&A, 620, A167, 2018, reproduced with permission ©ESO

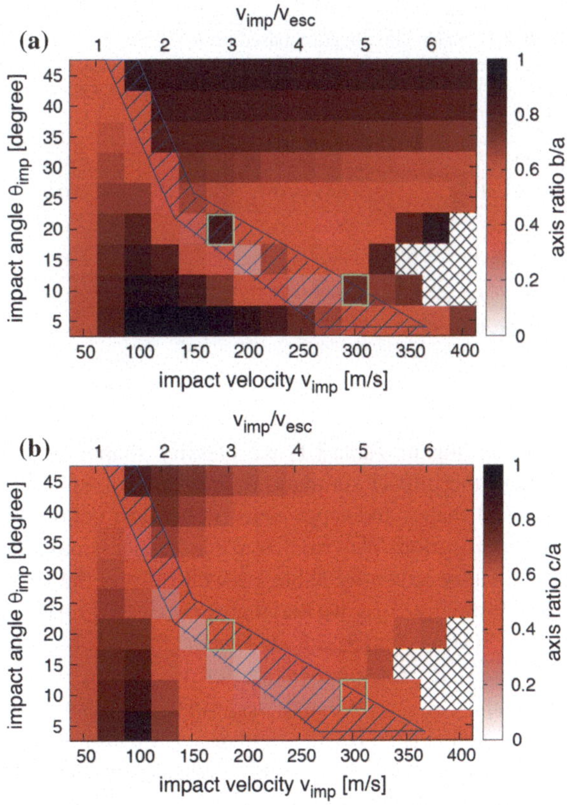

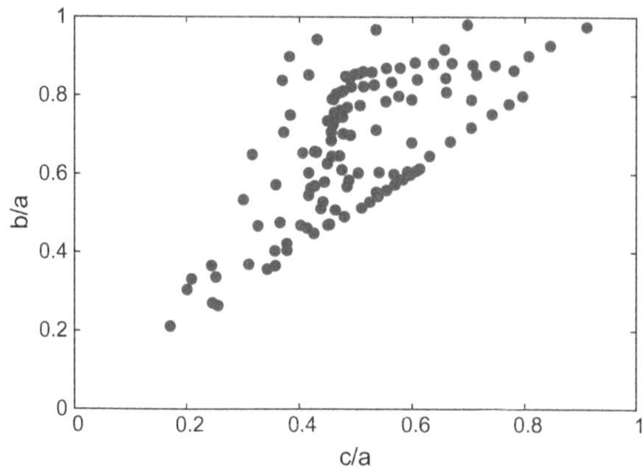

Fig. 3.15 Axis ratios of the largest remnants with $M_{\mathrm{lr}} > 0.4 M_{\mathrm{target}}$ produced through equal-mass and non-destructive impacts. The horizontal axis represents the ratio c/a, and the vertical axis represents the ratio b/a. Each point shows the axis ratios of the largest remnant produced through a non-destructive impact

Table 3.1 Shape classification based on b/a, c/a, $M_{\mathrm{lr}}/M_{\mathrm{target}}$, and $M_{\mathrm{lr}}/M_{\mathrm{tot}}$ of the largest remnants, where $M_{\mathrm{tot}} = M_{\mathrm{target}} + M_i$ shows the total mass in each impact simulation. Sugiura et al., A&A, 620, A167, 2018, reproduced with permission ©ESO

Shape	Thresholds		
Bilobed	$b/a < 0.7$	$c/a < 0.7$	$M_{\mathrm{lr}}/M_{\mathrm{tot}} = 1.0$
Spherical	$b/a > 0.7$	$c/a > 0.7$	...
Flat	$b/a > 0.7$	$c/a < 0.7$	$M_{\mathrm{lr}}/M_{\mathrm{target}} > 1.0$
Elongated	$b/a < 0.7$	$c/a < 0.7$	$M_{\mathrm{lr}}/M_{\mathrm{tot}} < 1.0$
Hemispherical	$b/a > 0.7$	$c/a < 0.7$	$M_{\mathrm{lr}}/M_{\mathrm{target}} < 1.0$
Super-catastrophic	...	...	$M_{\mathrm{lr}}/M_{\mathrm{target}} < 0.4$

According to Table 3.1, we classify shapes of the largest remnants produced through the impact simulations into bilobed, spherical, flat, elongated, and hemispherical shapes and super-catastrophic destruction. Figure 3.16 shows the result of the classification. We visualize some impact simulations and confirm that the shape classification given by Table 3.1 well reflects the shapes resulted from the simulations. From Fig. 3.16, we find that we need the impact conditions of $v_{\mathrm{imp}} \sim 50\,\mathrm{m/s}$ or $v_{\mathrm{imp}} \sim 100\,\mathrm{m/s}$ and $\theta_{\mathrm{imp}} > 25°$ for the formation of bilobed shapes, $v_{\mathrm{imp}} \sim 100\,\mathrm{m/s}$ and $\theta_{\mathrm{imp}} < 25°$ for spherical shapes, $v_{\mathrm{imp}} > 100\,\mathrm{m/s}$ and $v_{\mathrm{imp}} \sin\theta_{\mathrm{imp}} < 30\,\mathrm{m/s}$ for flat shapes, $v_{\mathrm{imp}} > 100\,\mathrm{m/s}$, $v_{\mathrm{imp}} \sin\theta_{\mathrm{imp}} > 30\,\mathrm{m/s}$, and $\theta_{\mathrm{imp}} < 30°$ for elongated shapes, and $v_{\mathrm{imp}} > 100\,\mathrm{m/s}$ and $\theta_{\mathrm{imp}} > 30°$ for hemispherical shapes.

The two impacts ($v_{\mathrm{imp}} = 175\,\mathrm{m/s}$, $\theta_{\mathrm{imp}} = 20°$ and $v_{\mathrm{imp}} = 300\,\mathrm{m/s}$, $\theta_{\mathrm{imp}} = 10°$) produce flat shapes even though they are in the parameter region that results in the formation of elongated shapes. These impacts correspond to elongated-shape-forming collisions with the second collision (see Fig. 3.13). As discussed above,

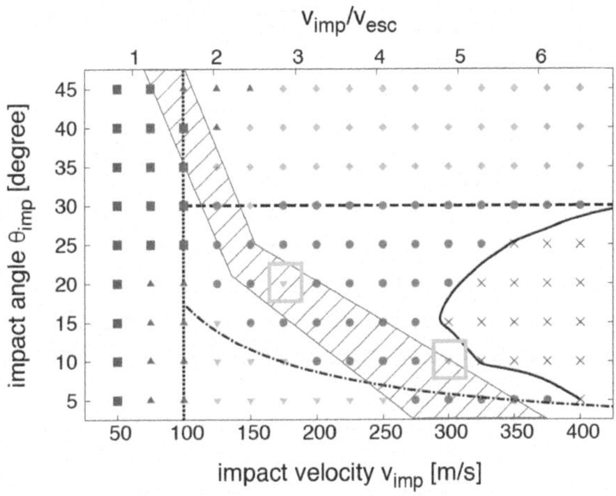

Fig. 3.16 Result of the shape classification of the largest remnants produced through various impacts with various v_{imp} (horizontal axis) and θ_{imp} (vertical axis). The blue squares represent bilobed shapes, the gray triangles represent spherical shapes, the orange inverted triangles represent flat shapes, the red circles represent elongated shapes, the light green diamonds represent hemispherical shapes, and the black crosses represent super-catastrophic destruction. The dotted line shows $v_{\mathrm{imp}} = 100\,\mathrm{m/s} \approx 1.6v_{\mathrm{esc}}$, the dashed line shows $\theta_{\mathrm{imp}} = 30^\circ$, the chain curve shows $v_{\mathrm{imp}} \sin\theta_{\mathrm{imp}} = 30\,\mathrm{m/s} \approx 0.5v_{\mathrm{esc}}$, and the solid curve shows the contour of $M_{\mathrm{lr}} = 0.4M_{\mathrm{target}}$. The hatched region and green boxes show the same meanings as in Figs. 3.5 and 3.14. Sugiura et al., A&A, 620, A167, 2018, reproduced with permission ©ESO

these impacts can be considered as elongated-shape-forming collisions because the shapes after the first collisions are elongated.

Based on Fig. 3.16, we summarize the required impact conditions for the formation of each shape as follows:

- bilobed and spherical shapes: $v_{\mathrm{imp}} < 1.6v_{\mathrm{esc}}$,
- flat shapes: $v_{\mathrm{imp}} > 1.6v_{\mathrm{esc}}$ and $v_{\mathrm{imp}} \sin\theta_{\mathrm{imp}} < 0.5v_{\mathrm{esc}}$,
- hemispherical shapes: $v_{\mathrm{imp}} > 1.6v_{\mathrm{esc}}$ and $\theta_{\mathrm{imp}} > 30^\circ$,

and

- elongated shapes: $v_{\mathrm{imp}} > 1.6v_{\mathrm{esc}}$, $\theta_{\mathrm{imp}} < 30^\circ$, $v_{\mathrm{imp}} \sin\theta_{\mathrm{imp}} > 0.5v_{\mathrm{esc}}$, and $M_{\mathrm{lr}} > 0.4M_{\mathrm{target}}$,

where $v_{\mathrm{esc}} \approx 60\,\mathrm{m/s}$ shows the two-body escape velocity. We note that the elongated-shape-forming region in Fig. 3.16 is separated by the hatched region into left and right parts. Elongated shapes in the left part are formed through merging collisions (see Fig. 3.2), while those in the right part are formed through hit-and-run collisions (see Fig. 3.9).

3.3.5 Discussion: Thresholds to Distinguish Impacts that Form Each Shape

For $v_{imp} < 1.6v_{esc}$, the degree of deformation is low, and spherical and bilobed shapes are formed. In contrast, elongated and flat shapes, i.e., highly deformed shapes, are formed for $v_{imp} > 1.6v_{esc}$. Thus the threshold $v_{imp} = 1.6v_{esc}$ represents the condition for significant deformation. The condition is theoretically proved as follows. During impacts and subsequent deformation, typical frictional force $\mu_d P$ with typical compression pressure P acts on the area of $\sim\pi R_t^2$ with typical deformation length of $\sim 4R_t$. The dissipated energy due to the frictional force E_{dis} during the deformation is then estimated as

$$E_{dis} = 4\pi R_t^3 \mu_d P. \tag{3.5}$$

Typical timescale for the deformation is $\sim 4R_t/v_{imp}$. The timescale is much longer than shock passing time $\sim 4R_t/C_s$ with the sound speed $C_s \approx 3\,\mathrm{km/s}$. Shock waves caused by impacts produce high pressure states, but they are relaxed within the shock passing time. Thus shocks do not contribute to the compression pressure for frictional energy dissipation, and the self-gravity mainly contributes to the compression pressure even during impact deformation. The typical compression pressure due to the self-gravity is estimated to be central pressure of an asteroid, which is given by

$$P = \frac{2}{3}\pi G \rho_0^2 R_t^2. \tag{3.6}$$

Equating the dissipated energy due to the friction E_{dis} to the total kinetic energy for equal-mass impacts $(1/4)M_{target}v_{imp}^2$, the impact velocity required for significant deformation is estimated as follows:

$$\begin{aligned} v_{imp,crit} &= \sqrt{4E_{dis}/M_{target}} = \sqrt{3\mu_d}\, v_{esc} \\ &= 1.587\left(\frac{\mu_d}{\tan(40°)}\right)^{1/2} v_{esc}. \end{aligned} \tag{3.7}$$

Although the above estimation uses crude approximation, the impact velocity obtained in Eq. (3.7) well agrees with the threshold $v_{imp} = 1.6v_{esc}$.

The threshold $\theta_{imp} = 30°$ determines whether sufficient parts of bodies directly collide or not. Only half or smaller portion of a body is intersected by another body for impacts with $\theta_{imp} \geq 30°$, which merely results in hit-and-run collisions and the formation of hemispherical shapes (Asphaug 2010; Leinhardt and Stewart 2012). In contrast, almost whole volume of two impacting bodies is deformed and elongated for impacts with $\theta_{imp} < 30°$, which results in the formation of elongated shapes.

Elongated shapes are formed through stretch of asteroids in the direction of shear velocity (see Figs. 3.2 and 3.9), and thus sufficient shear velocity of $v_{imp}\sin\theta_{imp} > 0.5v_{esc}$ is also necessary for the formation of elongated shapes. In contrast, low shear velocity of $v_{imp}\sin\theta_{imp} < 0.5v_{esc}$ results in the formation of flat shapes. We find that the self-gravity prevents deformation and thus elongation needs $v_{\rm imp}\sin\theta_{\rm imp} > v_{\rm esc}/2$.

Catastrophic collisions mainly produce spherical shapes although elongated or bilobed shapes are also formed through catastrophic collisions (see Fig. 3.12 or Sect. 3.5). For certain formation of elongated shapes, we need the condition of $M_{\rm lr} < 0.4M_{\rm target}$.

3.4 Unequal-Mass and Non-destructive Impacts

In this section, we introduce asteroidal shapes formed through non-destructive impacts, but we consider impacts with mass of impactor asteroids that is different from that of target asteroids. We again focus on non-destructive impacts with $M_{\rm lr} > 0.4M_{\rm target}$ and investigate shapes of the largest remnants. Except for impact simulations with the mass ratio $q = 1/64$, the number of SPH particles for target asteroids is fixed to 50,000, which is the same resolution as that in Sect. 3.3. For the simulations with $q = 1/64$, we use 100,000 SPH particles for target asteroids to use sufficient number of SPH particles for impactor asteroids.

3.4.1 Similar-Mass Impacts

Firstly, we introduce simulation results of impacts between two asteroids that have relatively similar masses. We conduct the simulations with the mass ratios of $q = 0.75$, 0.50, and 0.25. We vary the impact angle from 5° to 45° with the increment of 5°. We also vary the impact velocity from 50 to 600 m/s with the increment of 50 m/s, which is wider range but larger increment of the impact velocity compared to the parameter survey for the equal-mass impacts in Sect. 3.3.

Figure 3.17 shows the summary of the mass and axis ratios of the largest remnants. From the panels of $M_{\rm lr}/M_{\rm target}$ in Fig. 3.17, we observe clear boundaries between merging collisions and erosive/hit-and-run collisions for all of the three mass ratios. We also notice that the impacts with $v_{\rm imp} = 600$ m/s and $\theta_{\rm imp} = 5°$ result in $M_{\rm lr} < 0.4M_{\rm target}$, i.e., catastrophic disruption, even with $q = 0.25$. Thus the impacts with 50 m/s $\leq v_{\rm imp} \leq$ 600 m/s include various types of collisions, namely, merging collisions, hit-and-run collisions, and catastrophic disruptions.

From panels of b/a and c/a in Fig. 3.17, we observe an entire tendency that the axis ratios increase with decreasing the mass ratios. This means that the impacts with high-mass ratios are difficult to produce irregular shapes. Especially, this tendency is clear for b/a; $b/a \approx 1$ for almost all of the impacts with $q = 0.25$. Thus, non-destructive impacts with high-mass ratios do not produce elongated shapes of the largest remnants.

The left panels of Fig. 3.18 show shape classification of the largest remnants for similar-mass impacts. We use the same threshold in Table 3.1 for the classification as

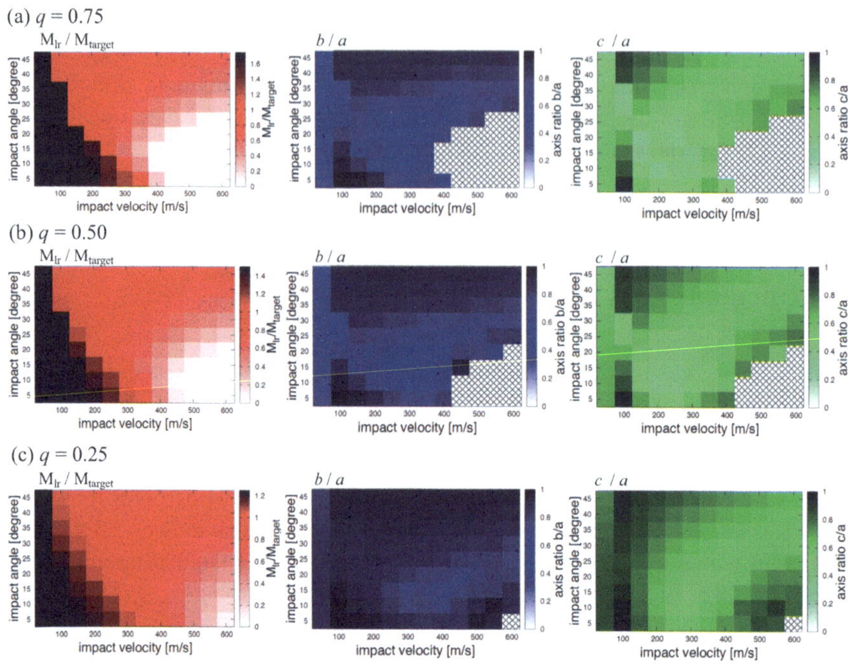

Fig. 3.17 Summary of the mass and axis ratios of the largest remnants produced through the impacts with three different mass ratios. Each row shows the results of the impacts with **a** $q = 0.75$, **b** $q = 0.50$, and **c** $q = 0.25$. Each color of contour shows $M_{\mathrm{lr}}/M_{\mathrm{target}}$ (red), b/a (blue), and c/a (green) of the largest remnants. In the panels that show $M_{\mathrm{lr}}/M_{\mathrm{target}}$, black regions show the impacts with $M_{\mathrm{lr}} \approx M_{\mathrm{tot}}$, i.e., the impacts resulting in merging collisions. In each panel, the horizontal axis represents the impact velocity, and the vertical axis represents the impact angle. The impacts in cross-hatched region have $M_{\mathrm{lr}} < 0.15 M_{\mathrm{target}}$, and we do not measure the axis ratios

that for the equal-mass impacts. From Fig. 3.18a, b, we notice that the results of the classification for $q = 0.75$ and $q = 0.50$ are similar to that of the equal-mass impacts (Fig. 3.16): In the impact velocity-angle diagram, the left parts are for bilobed shapes, the left-bottom parts are for spherical shapes, the bottom parts are for flat shapes, the middle parts are for elongated shapes, the upper parts are for hemispherical shapes, and the right parts are for catastrophic disruptions.[1] However, the result of the shape classification for $q = 0.25$ is distinctly different from that of the equal-mass impacts. There is no clear region for bilobed shapes and elongated shapes. This is because the impacts with high-mass ratios tend to result in partial destruction rather than overall deformation. Thus the various shapes of the largest remnants are only formed through the impacts with $q \geq 0.5$.

[1] We notice that there are some exceptions, but these exceptions are mainly caused by the thresholds we determined for the equal-mass impacts. If we could find better thresholds, we might remove the exceptions.

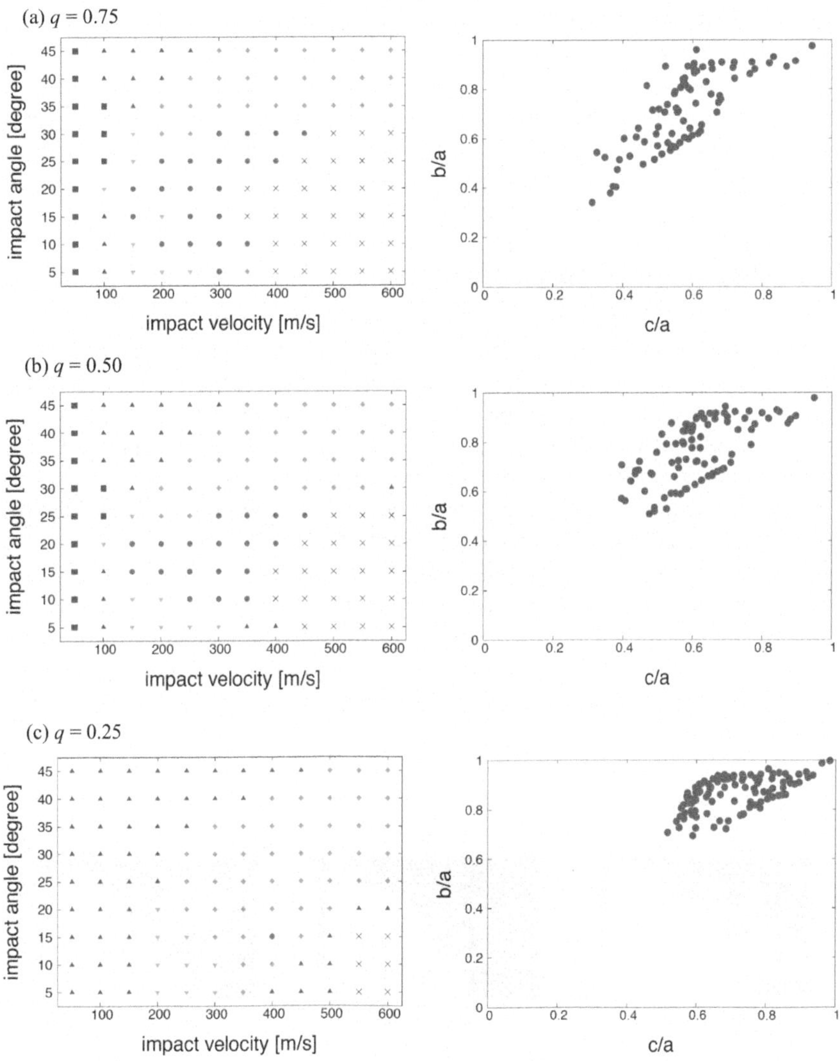

Fig. 3.18 Summary of the classification of resultant shapes and the shape distribution of the largest remnants produced through the impacts with the three different mass ratios. Each row shows the results of the impacts with **a** $q = 0.75$, **b** $q = 0.50$, and **c** $q = 0.25$. The left panels show the results of the shape classification according to Table 3.1 with the horizontal axis of the impact velocity and the vertical axis of the impact angle. The meanings of symbols are the same as those of Fig. 3.16. The right panels show the axis ratios of the largest remnants with the horizontal axis of c/a and the vertical axis of b/a

The right panels of Fig. 3.18 show the shape distribution of the largest remnants produced through the similar-mass and non-destructive impacts with $M_{\rm lr} > 0.4M_{\rm target}$. From these panels, we clearly observe that the number of elongated shapes of impact outcomes decreases with decreasing the mass ratio q. Especially for $q = 0.25$, even flat shapes with $c/a \approx 0.5$ are not found. Thus, to produce extreme shapes, we need non-destructive impacts with $q \geq 0.5$.

3.4.2 Impacts with Large Mass Ratios

Secondly, we introduce results of collisions between two asteroids that have large mass differences. We investigate asteroidal collisions with the mass ratios of $q = 1/8$ and $1/64$, i.e., even the ratio of the radius R_i/R_t is smaller than 0.5. For $q = 1/8$, we vary the impact velocity from 125 to 2,500 m/s with the increment of 125 m/s and the impact angle from 0° to 45° with the increment of 5°. For $q = 1/64$, we vary the impact velocity from 500 to 3,000 m/s with the increment of 250 m/s and the impact angle from 0° to 40° with the increment of 10°.

Figure 3.19 summarizes the mass and the axis ratios of the largest remnants produced through the high-mass-ratio impacts. From the panels of $M_{\rm lr}/M_{\rm target}$ in Fig. 3.19, we notice that our investigation of the high-mass-ratio impacts covers various impacts from non-destructive impacts to destructive impacts especially for $q = 1/8$. The parameter survey for $q = 1/64$ also includes destructive impacts with $M_{\rm lr} < 0.4M_{\rm target}$.

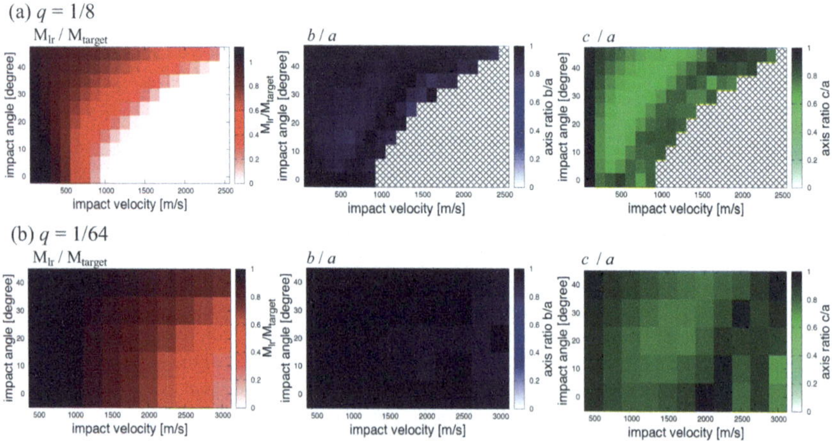

Fig. 3.19 Summary of the mass and axis ratios of the largest remnants produced through the high-mass-ratio impacts. Each row shows the results of the impacts with **a** $q = 1/8$ and **b** $q = 1/64$. Each color of contour shows $M_{\rm lr}/M_{\rm target}$ (red), b/a (blue), and c/a (green) of the largest remnants. In each panel, the horizontal axis represents the impact velocity, and the vertical axis represents the impact angle. The impacts in cross-hatched region have $M_{\rm lr} < 0.15M_{\rm target}$, and we do not measure the axis ratios

Figure 3.19 shows that shapes of the largest remnants produced through the high-mass-ratio impacts are only spherical and hemispherical shapes. The minimum b/a of the largest remnants produced through the impacts with $q = 1/8$ is about 0.8. All largest remnants produced through the impacts with $q = 1/64$ have $b/a > 0.9$. Thus, high-mass-ratio and non-destructive impacts do not produce elongated shapes.

From the panels of c/a in Fig. 3.19, we also notice a tendency that c/a increases with decreasing q. The minimum c/a resulted from the impacts with $q = 1/8$ is about 0.6, while that resulted from the impacts with $q = 1/64$ is about 0.7. This is because the high-mass-ratio and non-destructive impacts mainly result in cratering. The high-mass-ratio impacts do not produce largely irregular shapes even if the impacts result in relatively destructive impacts with $M_{\rm lr} \approx 0.5 M_{\rm target}$, because such impacts mainly eject materials around impact cites rather than overall deformation. Therefore, our simulations suggest that high-mass-ratio and non-destructive impacts with $q \leq 1/64$ do not produce irregular shapes of the largest remnants.

3.5 Destructive Impacts

In this section, we introduce asteroidal shapes formed through destructive impacts. To investigate shapes of small remnants produced through destructive impacts, we conduct high-resolution simulations of catastrophic destruction. We use about 4 million SPH particles for each impact simulation and conduct four simulations with the following conditions: $q = 1$, $\theta_{\rm imp} = 15°$, $v_{\rm imp} = 350\,{\rm m/s}$; $q = 1/4$, $\theta_{\rm imp} = 15°$, $v_{\rm imp} = 700\,{\rm m/s}$; $q = 1/8$, $\theta_{\rm imp} = 0°$, $v_{\rm imp} = 1{,}000\,{\rm m/s}$; and $q = 1/16$, $\theta_{\rm imp} = 15°$, $v_{\rm imp} = 1{,}700\,{\rm m/s}$.

Figure 3.20 shows snapshots of the impacts with $q = 1$, $\theta_{\rm imp} = 15°$, $v_{\rm imp} = 350\,{\rm m/s}$ and $q = 1/8$, $\theta_{\rm imp} = 0°$, $v_{\rm imp} = 1{,}000\,{\rm m/s}$. We notice that the catastrophic destruction produces large ejecta curtains regardless of the detailed impact conditions such as the mass ratio (Fig. 3.20a-2, c-2). The destructive impacts produce thin ejecta curtains (Fig. 3.20b-2, d-2), and the gravitational instability and fragmentation of the curtains produce many filamentary structures (Fig. 3.20a-4, c-4). The gravitational reaccumulation of fragments in the filamentary structures (Fig. 3.20a-5, c-5) eventually leads to the formation of many small remnants (Fig. 3.20a-6, c-6). Note that these processes do not depend on detailed impact conditions as long as impacts result in catastrophic destruction.

Figure 3.21 shows the cumulative mass distributions of the remnants produced through the four destructive impacts. From Fig. 3.21, we notice that the masses of the largest remnants produced through the impacts with $q = 1$, $\theta_{\rm imp} = 15°$ $v_{\rm imp} = 350\,{\rm m/s}$; $q = 1/4$, $\theta_{\rm imp} = 15°$, $v_{\rm imp} = 700\,{\rm m/s}$; and $q = 1/16$, $\theta_{\rm imp} = 15°$, $v_{\rm imp} = 1{,}700\,{\rm m/s}$ are about $0.1 M_{\rm target}$, while that with $q = 1/8$, $\theta_{\rm imp} = 0°$, $v_{\rm imp} = 1{,}000\,{\rm m/s}$ is about $0.01 M_{\rm target}$. We also notice that the ratios of the mass of the largest to that of the second largest for the impacts with $q = 1$, $\theta_{\rm imp} = 15°$, $v_{\rm imp} = 350\,{\rm m/s}$ and $q = 1/8$, $\theta_{\rm imp} = 0°$, $v_{\rm imp} = 1{,}000\,{\rm m/s}$ are about unity, while those with $q = 1/4$, $\theta_{\rm imp} = 15°$, $v_{\rm imp} = 700\,{\rm m/s}$ and $q = 1/16$, $\theta_{\rm imp} = 15°$, $v_{\rm imp} = 1{,}700\,{\rm m/s}$ are less

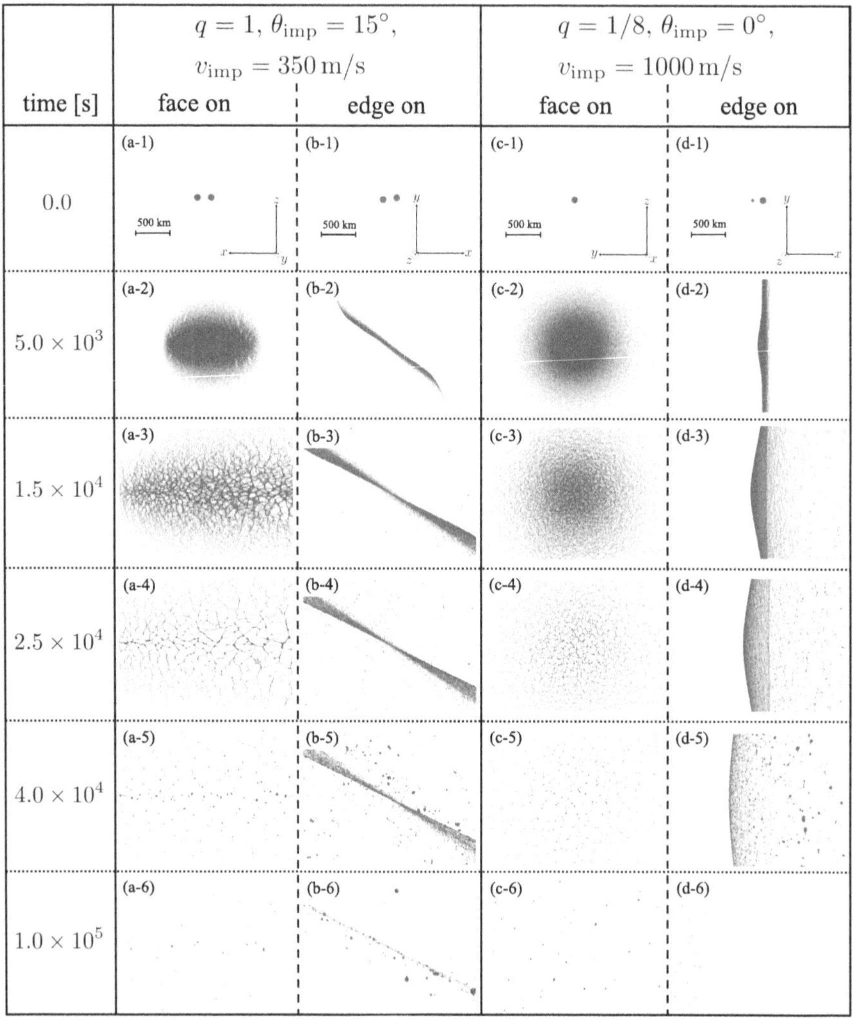

Fig. 3.20 Snapshots of the simulations of the destructive impacts. The panels (a-1)–(a-6) and (b-1)–(b-6) show the face on views and the edge on views of the ejecta curtain produced through the impact with $q = 1$, $\theta_{\rm imp} = 15°$, $v_{\rm imp} = 350\,{\rm m/s}$, respectively. The panels (c-1)–(c-6) and (d-1)–(d-6) show the face on views and the edge on views of the ejecta curtain produced through the impact with $q = 1/8$, $\theta_{\rm imp} = 0°$, $v_{\rm imp} = 1{,}000\,{\rm m/s}$, respectively. Each row shows the snapshots at 0.0s, 5.0×10^3s, 1.5×10^4s, 2.5×10^4s, 4.0×10^4s, and, 1.0×10^5s. In the panels (a-1), (b-1), (c-1), and (d-1), we show the scales and the coordinate systems that are valid for each column

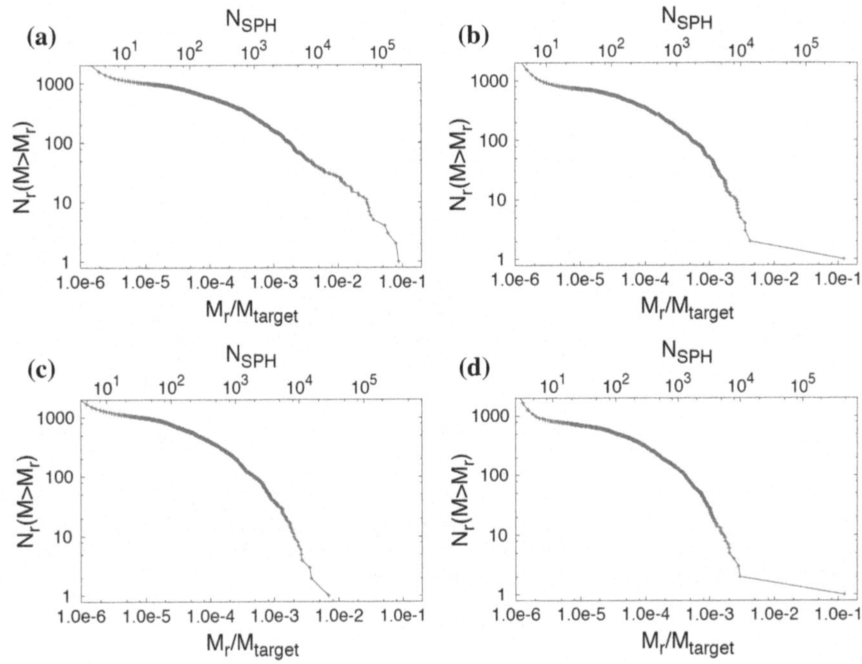

Fig. 3.21 Cumulative mass distributions of the remnants produced through the destructive impacts. In each panel, the horizontal axis represents the mass of remnants M_r normalized by the mass of the target asteroids $M_{\rm target}$, and the vertical axis represents the cumulative number of remnants that have the mass larger than M_r. The upper horizontal axis represents the number of SPH particles that compose a remnant with the mass of M_r. The panel **a** shows the cumulative mass distribution resulted from the impact with $q = 1$, $\theta_{\rm imp} = 15°$, $v_{\rm imp} = 350\,{\rm m/s}$, the panel **b** shows that with $q = 1/4$, $\theta_{\rm imp} = 15°$, $v_{\rm imp} = 700\,{\rm m/s}$, the panel **c** shows that with $q = 1/8$, $\theta_{\rm imp} = 0°$, $v_{\rm imp} = 1{,}000\,{\rm m/s}$, and the panel **d** shows that with $q = 1/16$, $\theta_{\rm imp} = 15°$, $v_{\rm imp} = 1{,}700\,{\rm m/s}$

than 0.1. We do not investigate the dependence of the feature of cumulative mass distributions on the impact conditions in more detail, but it should be noted that not only the mass of the largest remnants but also the ratios of the mass of the largest to the second largest probably have important information to reveal impact conditions that formed asteroid families.

If a remnant is composed of 5,000 SPH particles, this remnant is resolved by about 20 SPH particles along axis directions. This number of SPH particles is considered to be sufficient to measure the shape of such remnant. Thus, we measure the axis ratios of remnants composed of 5,000 or more SPH particles. From Fig. 3.21, we notice that the number of the remnants with 5,000 or more SPH particles for the impacts with $q = 1$, $\theta_{\rm imp} = 15°$ $v_{\rm imp} = 350\,{\rm m/s}$ is 69, that with $q = 1/4$, $\theta_{\rm imp} = 15°$, $v_{\rm imp} = 700\,{\rm m/s}$ is 24, that with $q = 1/8$, $\theta_{\rm imp} = 0°$, $v_{\rm imp} = 1{,}000\,{\rm m/s}$ is 30, and that with $q = 1/16$, $\theta_{\rm imp} = 15°$, $v_{\rm imp} = 1{,}700\,{\rm m/s}$ is 13. Thus the total number of

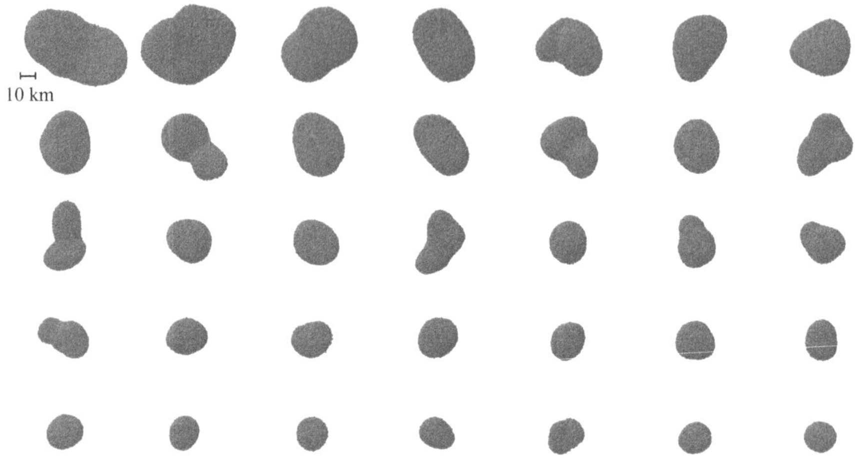

Fig. 3.22 Snapshots of the remnants produced through the impact with $q = 1$, $\theta_{\mathrm{imp}} = 15^\circ$, $v_{\mathrm{imp}} = 350\,\mathrm{m/s}$. We show the shapes of from the largest remnants (the remnant at the left upper corner) to the 35th largest remnant (the remnant at the right bottom corner). Reprinted from Sugiura et al. (2020), Copyright (2019), with permission from Elsevier

the remnants with measured shapes is sufficient to statistically discuss the shape distribution of remnants produced through destructive impacts.

Figure 3.22 shows snapshots of the main remnants produced through the impact with $q = 1$, $\theta_{\mathrm{imp}} = 15^\circ$ $v_{\mathrm{imp}} = 350\,\mathrm{m/s}$. From Fig. 3.22, we notice that the most remnants have rounded shapes rather than irregular shapes. This is because the remnants are produced through violent reaccumulation of fragments and gravity tends to produce spherical bodies. However, we also notice that some remnants have characteristic shapes with two lobes. These bilobed shapes are formed through coalescence of two rounded remnants. Figure 3.22 shows a tendency that the smaller remnants have more rounded shapes. This may be caused by the weak mutual gravity between the smaller remnants; they do not coalesce each other due to the weak mutual gravity, which leads to let these remnants remain spherical shapes.

Figure 3.23 shows the dependence of $\sqrt{bc}/a$ on the mass of the remnants. As shown in Schwartz et al. (2018), $\sqrt{bc}/a$ represents shape irregularity of remnants, and smaller $\sqrt{bc}/a$ means more irregular shapes. From Fig. 3.23, we notice a tendency that the smaller remnants have more rounded shapes for all the impacts. Schwartz et al. (2018) also investigate the shape distribution of remnants produced through several catastrophic impacts through simulations using a particle based code of a soft sphere model. However, they do not observe any tendency of $\sqrt{bc}/a$ on the mass of the remnants. This discrepancy may be caused by the difference of simulation methods. Thus, in our future works, we will investigate detailed dependence of shape distribution of produced remnants on simulation methods.

Figure 3.24 shows the shape distribution of the remnants produced through the destructive impacts. From Fig. 3.24, we notice that all the destructive impacts mainly

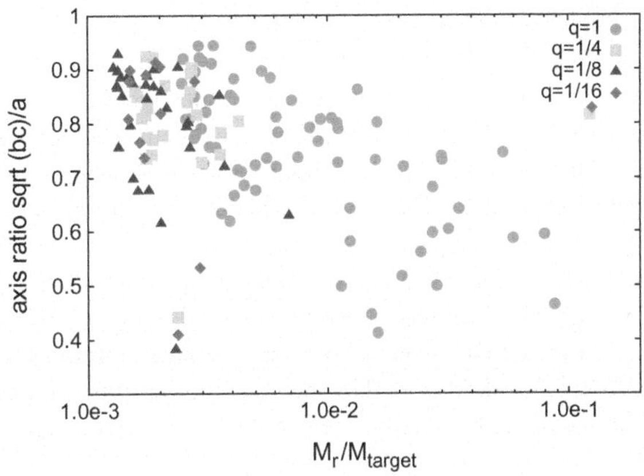

Fig. 3.23 Dependence of shape irregularity on the mass of the remnants. The horizontal axis represents the mass of the remnants normalized by that of the target asteroids, and the vertical axis represents $\sqrt{bc}/a$ of the remnants. The red circles show the remnants produced through the impact with $q = 1$, $\theta_{\rm imp} = 15°$, $v_{\rm imp} = 350\,{\rm m/s}$, the green squares show those with $q = 1/4$, $\theta_{\rm imp} = 15°$, $v_{\rm imp} = 700\,{\rm m/s}$, the blue triangles show those with $q = 1/8$, $\theta_{\rm imp} = 0°$, $v_{\rm imp} = 1{,}000\,{\rm m/s}$, and the brown diamonds show those with $q = 1/16$, $\theta_{\rm imp} = 15°$, $v_{\rm imp} = 1{,}700\,{\rm m/s}$. Reprinted from Sugiura et al. (2020), Copyright (2019), with permission from Elsevier

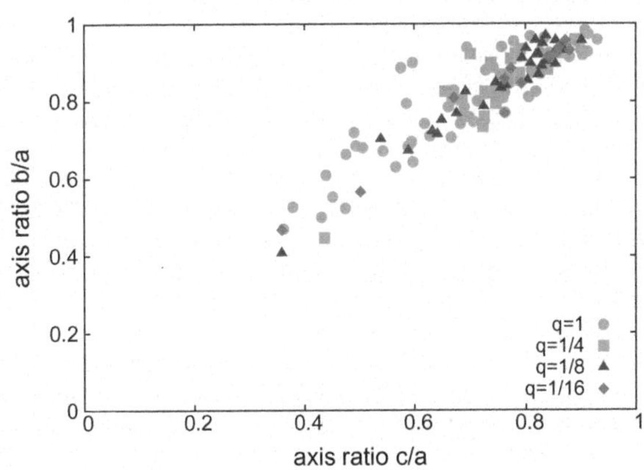

Fig. 3.24 Shape distribution of the remnants produced through the destructive impacts. The horizontal axis represents c/a of the remnants, and the vertical axis represents b/a of the remnants. The red circles show the remnants produced through the impact with $q = 1$, $\theta_{\rm imp} = 15°$, $v_{\rm imp} = 350\,{\rm m/s}$, the green squares show those with $q = 1/4$, $\theta_{\rm imp} = 15°$, $v_{\rm imp} = 700\,{\rm m/s}$, the blue triangles show those with $q = 1/8$, $\theta_{\rm imp} = 0°$, $v_{\rm imp} = 1{,}000\,{\rm m/s}$, and the brown diamonds show those with $q = 1/16$, $\theta_{\rm imp} = 15°$, $v_{\rm imp} = 1{,}700\,{\rm m/s}$. Reprinted from Sugiura et al. (2020), Copyright (2019), with permission from Elsevier

produce spherical shapes and also irregular shapes with $b/a \approx c/a \approx 0.5$ that are mainly bilobed shapes. We also notice that the destructive impacts are difficult to produce flat shapes. Among the remnants produced through the four impacts investigated here, there are only two flat shapes with $b/a > 0.8$ and $c/a < 0.6$, and these flat shapes are produced by the equal-mass and relatively low-velocity impact. Therefore, destructive impacts, especially high-velocity impacts, do not produce flat shapes of asteroids.

In this section, we investigated shapes of small remnants produce through destructive impacts. However, small remnants are also ejected by relatively non-destructive impacts (see, e.g., Fig. 3.11b). Shapes of fine remnants produced through non-destructive impacts are also determined by the reaccumulation of fragments. Thus we expect that the shapes of such remnants resemble those formed through destructive impacts.

References

C. Agnor, E. Asphaug, Accretion efficiency during planetary collisions. Astrophys. J. **613**, L157–L160 (2004). https://doi.org/10.1086/425158

E. Asphaug, Similar-sized collisions and the diversity of planets. Chem. Erde. Geochem. **70**, 199–219 (2010). https://doi.org/10.1016/j.chemer.2010.01.004

A. Fujiwara, G. Kamimoto, A. Tsukamoto, Expected shape distribution of asteroids obtained from laboratory impact experiments. Nature **272**, 602 (1978). https://doi.org/10.1038/272602a0

A. Fujiwara, J. Kawaguchi, D.K. Yeomans, M. Abe, T. Mukai, T. Okada, J. Saito, H. Yano, M. Yoshikawa, D.J. Scheeres, O. Barnouin-Jha, A.F. Cheng, H. Demura, R.W. Gaskell, N. Hirata, H. Ikeda, T. Kominato, H. Miyamoto, A.M. Nakamura, R. Nakamura, S. Sasaki, K. Uesugi, The rubble-pile asteroid Itokawa as observed by Hayabusa. Science **312**, 1330–1334 (2006). https://doi.org/10.1126/science.1125841

G.H. Heiken, D.T. Vaniman, B.M. French, *Lunar Sourcebook—A User's Guide to the Moon* (1991)

J.P. Huchra, M.J. Geller, Groups of galaxies. I—Nearby groups. Astrophys. J. **257**, 423–437 (1982). https://doi.org/10.1086/160000

Z.M. Leinhardt, S.T. Stewart, Collisions between gravity-dominated bodies. I. Outcome regimes and scaling laws. Astrophys. J. **745**, 79 (2012). https://doi.org/10.1088/0004-637X/745/1/79

T. Michikami, T. Kadokawa, A. Tsuchiyama, A. Hagermann, T. Nakano, K. Uesugi, S. Hasegawa, Influence of petrographic textures on the shapes of impact experiment fine fragments measuring several tens of microns: comparison with Itokawa regolith particles. Icarus **302**, 109–125 (2018). https://doi.org/10.1016/j.icarus.2017.10.040

S.R. Schwartz, P. Michel, M. Jutzi, S. Marchi, Y. Zhang, D.C. Richardson, Catastrophic disruptions as the origin of bilobate comets. Nat. Astron. **2**, 379–382 (2018). https://doi.org/10.1038/s41550-018-0395-2

K. Sugiura, H. Kobayashi, S. Inutsuka, Toward understanding the origin of asteroid geometries. Variety in shapes produced by equal-mass impacts. Astron. Astrophys. **620**, A167 (2018). https://doi.org/10.1051/0004-6361/201833227

K. Sugiura, H. Kobayashi, S. Inutsuka, High-resolution simulations of catastrophic disruptions: resultant shape distributions. Planet. Space Sci. **181**, 104807 (2020). https://doi.org/10.1016/j.pss.2019.104807

Chapter 4
Discussion: Collisional History of Asteroids

Abstract We found that low-velocity and non-destructive impacts produce various shapes including extremely flat shapes, while high-velocity and destructive impacts produce spherical and bilobed shapes but they do not produce flat shapes. The impact velocity is strongly affected by the environments of planetary systems such as the existence of gas components or giant planets. Low-velocity impacts are realized in the primordial environment, while high-velocity impacts are realized in the present environment. Thus we may connect shapes of asteroids to the environments or era when asteroids were formed. In this chapter, we discuss the connection between shapes of asteroids and formation era. We also investigate information of actual asteroids including those shapes obtained from several databases, and we find that many flat asteroids are not included in any asteroid families. Both the results of our simulations and the information of actual asteroids suggest that flat shapes of large asteroids are likely to be formed in the primordial environment and they keep their flat shapes until today.

Keywords Asteroids · Shapes · Formation era

4.1 Connection Between Shapes of Asteroids and Collisional Environments

4.1.1 Summary of Shapes of Impact Outcomes

In Chap. 3, we investigated asteroidal shapes produced through impacts with the 100 km target asteroids. We classified impacts into three categories; equal-mass and non-destructive impacts, unequal-mass and non-destructive impacts, and destructive impacts. Non-destructive impacts mean impacts with the mass of the largest remnants larger than $0.4M_{\mathrm{target}}$. We briefly introduce shapes of asteroids produced through the three types of impacts.

Equal-mass and non-destructive impacts with $v_{\mathrm{imp}} \leq 400\,\mathrm{m/s}$ produce various shapes including extremely elongated shapes (Fig. 3.2) and extremely flat shapes

K. Sugiura, *Development of a Numerical Simulation Method for Rocky Body Impacts and Theoretical Analysis of Asteroidal Shapes*, Springer Theses,
https://doi.org/10.1007/978-981-15-3722-6_4

(Fig. 3.8). The shapes of impact outcomes depend on impact conditions. Elongated shapes are formed through impacts with $2v_{\rm esc} \lesssim v_{\rm imp} \lesssim 5v_{\rm esc}$ and $15° \lesssim \theta_{\rm imp} \lesssim 30°$, and flat shapes are formed through impacts with $2v_{\rm esc} \lesssim v_{\rm imp} \lesssim 4v_{\rm esc}$ and $\theta_{\rm imp} \lesssim 15°$ (Fig. 3.16). The most elongated shape produced through the equal-mass and non-destructive impacts has $b/a \approx 0.2$ and the most flat shape has $c/a \approx 0.4$ (Fig. 3.15).

Unequal-mass and non-destructive impacts are more difficult to produce irregular shapes compared to equal-mass impacts. Impacts with $q = 0.5$ still produce elongated shapes with $b/a \approx 0.6$ (Fig. 3.18b) and those with $q = 0.25$ still produce flat shapes with $c/a \approx 0.6$ (Fig. 3.18c). However, the most irregular shape produced through impacts with $q = 1/8$ has $b/a \approx 1$ and $c/a \approx 0.6$ (Fig. 3.19a), and that produced through impacts with $q = 1/64$ has $b/a \approx 1$ and $c/a \approx 0.7$ (Fig. 3.19b). Thus, impacts with $q < 1/8$ do not produce irregular shapes with $c/a < 0.6$.

Destructive impacts mainly produce spherical and bilobed shapes (Fig. 3.22). Intense gravitational reaccumulation of fragments after destructive impacts produce spherical remnants, and the coalescence of spherical remnants produce bilobed shapes. We conduct simulations of destructive impacts with various mass ratios and impact velocities up to about 2 km/s and find that every destructive impact produces spherical and bilobed shapes (Fig. 3.24). Some bilobed remnants have $b/a \approx 0.4$. However, flat shapes are difficult to be formed through destructive impacts; among about 100 remnants produced through the destructive impacts, there are only two flat remnants with $c/a \approx 0.6$, and there are no flat remnants with $c/a \leq 0.5$.

4.1.2 *Impact Velocity and Collisional Environments*

4.1.2.1 Primordial Environment

We estimate typical relative velocity between asteroids in the primordial environment. We consider the era when protoplanets are under oligarchic growth, i.e., several protoplanets with sufficient separations grow through the accretion of surrounding asteroids. In this era, gas component of the protoplanetary disk still exists because Jupiter is not yet formed. Random velocity between asteroids increases through viscous stirring of protoplanets, while the random velocity decreases through gas drag.

Timescale of viscous stirring is estimated as (e.g., Ida and Makino 1993)

$$t_{\rm vis} = \frac{1}{n_{\rm proto}\sigma_{\rm scat} v_{\rm rel}} = \frac{1}{n_{\rm proto} v_{\rm rel}} \frac{1}{\pi R_{\rm proto}^2} \left(\frac{v_{\rm rel}}{v_{\rm esc,p}}\right)^4, \tag{4.1}$$

where $n_{\rm proto}$ is number density of protoplanets, $\sigma_{\rm scat} = \pi R_{\rm proto}^2 (v_{\rm esc,p}/v_{\rm rel})^4$ is scattering cross section of protoplanets, $R_{\rm proto}$ is radius of protoplanets, $v_{\rm esc,p}$ is escape velocity from protoplanets, and $v_{\rm rel}$ is the typical relative velocity between asteroids. We also denote mass of protoplanets as $M_{\rm proto} = (4/3)\pi R_{\rm proto}^3 \rho_s$, where ρ_s is solid density. The number density of protoplanets is estimated as

$$n_{\rm proto} = \frac{1}{2\pi a \cdot b_{\rm sep} \cdot h_{\rm ast}}, \quad (4.2)$$

where a is orbital semi-major axis, $b_{\rm sep}$ is orbital separation between protoplanets, and $h_{\rm ast}$ is scale height of asteroid disk. $h_{\rm ast}$ is estimated as $e_{\rm ast}a$, where $e_{\rm ast}$ is typical eccentricity of asteroids. Since $e_{\rm ast}v_{\rm K} = v_{\rm rel}$, where $v_{\rm K}$ is the Keplerian velocity, $h_{\rm ast}$ is written as

$$h_{\rm ast} = av_{\rm rel}/v_{\rm K}. \quad (4.3)$$

The orbital separations between protoplanets are maintained to be several times of mutual hill radius (e.g., Kokubo and Ida 2002), and thus $b_{\rm sep}$ is written as

$$b_{\rm sep} = \tilde{b}_{\rm sep}\left(\frac{2M_{\rm proto}}{3M_s}\right)^{1/3} a, \quad (4.4)$$

where $\tilde{b}_{\rm sep}$ is a factor and typically has $\tilde{b}_{\rm sep} \approx 10$, and M_s is the mass of the sun.

Timescale of the gas damping is estimated to be stopping time. For large asteroids with the radius $R_{\rm ast} = 50$ km, gas drag law is described by Newton law. The stopping time is written as (e.g., Whipple 1972; Weidenschilling 1977)

$$t_s = \frac{6\rho_s R_{\rm ast}}{\rho_{\rm gas} v_{\rm rel}}, \quad (4.5)$$

where $\rho_{\rm gas}$ is density of gas component.

Equating Eqs. (4.1) and (4.5), we achieve the relative velocity between asteroids at the equilibrium as

$$v_{\rm rel} = 300\left(\frac{R_{\rm ast}}{50\,{\rm km}}\right)^{1/5}\left(\frac{M_{\rm proto}}{M_{\rm Mars}}\right)^{1/3}\left(\frac{a}{3{\rm AU}}\right)^{-7/10} \times \left(\frac{\rho_{\rm gas}}{2\times10^{-9}\,{\rm g/cm^{-3}}}\right)^{-1/5}\left(\frac{\tilde{b}_{\rm sep}}{10}\right)^{-1/5}\left(\frac{\rho_s}{1\,{\rm g/cm^3}}\right)^{1/5} {\rm m/s}, \quad (4.6)$$

where $M_{\rm Mars}$ is the mass of Mars and $2 \times 10^{-9}\,{\rm g/cm^3}$ is the typical gas density obtained from the minimum mass solar nebula model (Hayashi 1981). Equation (4.6) shows that the relative velocity between asteroids is comparable to the escape velocity from asteroids with the radius 50 km $\approx$ 60 m/s until the mass of protoplanets grows to be Mars mass. Thus non-destructive impacts mainly occur when the planet formation is ongoing.

4.1.2.2 Present Environment

In the present solar system, typical impact velocity between asteroids is comparable to the Keplerian velocity because of the existence of the planets. The average relative velocity of asteroids is roughly estimated as $\sqrt{e_{\rm ave}^2 + i_{\rm ave}^2}\, v_{\rm K}$, where $e_{\rm ave}$ and $i_{\rm ave}$ are average orbital eccentricity and inclination, respectively. In the main belt, the Keplerian velocity is $v_{\rm K} \approx 20\,{\rm km/s}$. We achieve orbital elements of all observed asteroids from Jet Propulsion Laboratory (JPL) Small-Body Database,[1] and the average eccentricity and inclination are calculated as $e_{\rm ave} = 0.15$ and $i_{\rm ave} = 0.13$ (cf. Ueda et al. 2017). Thus, the average relative velocity is calculated as $\approx$4 km/s.

We also estimate the distribution of impact velocities between main-belt asteroids according to method for obtaining relative velocities at orbital crossing described in Whitmire et al. (1998) and Kobayashi and Ida (2001). We consider two asteroids 1 and 2 orbiting around the sun. Orbital elements of asteroid 1 and 2 have subscripts of 1 and 2, respectively. r, f, a, e, i, Ω, and ω represent orbital radius, true anomaly, semi-major axis, eccentricity, inclination, longitude of ascending node, and argument of pericenter, respectively, and $p = a(1 - e^2)$. $\Delta\tilde{\omega}$ shows the difference of longitude of pericenters of the two asteroids and is defined as $\Delta\tilde{\omega} = \tilde{\omega_2} - \tilde{\omega_1} = (\Omega_2 + \omega_2) - (\Omega_1 + \omega_1)$.

We assume the inclinations of the two asteroids are small and set $i_1 = i_2 = 0$, which leads to the simple condition of the orbital crossing of $r_1 = r_2$. Thus the orbital crossing occurs when

$$\left(\frac{1}{r_1} - \frac{1}{r_2}\right)_{f_2 = f_{2,c}} = 0 = \frac{1}{p_1} - \frac{1}{p_2} + \frac{e_1 \cos(f_{2,c} - \Delta\tilde{\omega})}{p_1} - \frac{e_2 \cos f_{2,c}}{p_2}, \tag{4.7}$$

where $f_{2,c}$ is true anomaly of the asteroid 2 at the orbital crossing. Note that we have the relation of $f_1 = f_2 - \Delta\tilde{\omega}$ when the two asteroids are located on the same line connecting the asteroids and the sun. The solution for $f_{2,c}$ of Eq. (4.7) is given by

$$\cos f_{2,c} = \frac{-AB \pm C\sqrt{C^2 + B^2 - A^2}}{B^2 + C^2}, \tag{4.8}$$

where

$$\begin{aligned} A &= p_2 - p_1, \\ B &= e_1 p_2 \cos \Delta\tilde{\omega} - e_2 p_1, \\ C &= e_1 p_2 \sin \Delta\tilde{\omega}. \end{aligned} \tag{4.9}$$

If the orbits of the two asteroids intersect each other, Eq. (4.8) has physical solutions, which requires

$$C^2 + B^2 - A^2 \geq 0. \tag{4.10}$$

Note that Eq. (4.8) has two solutions, and thus $f_{2,c}$ has two values if Eq. (4.10) is satisfied.

If Eq. (4.10) is satisfied, the orbits of the two asteroids are crossing under the assumption of $i_1 = i_2 = 0$. Then, we assess whether the two asteroids collide each

[1] https://ssd.jpl.nasa.gov/sbdb_query.cgi#x.

other or not in three-dimensional space when $f_2 = f_{2,c}$ and $f_1 = f_{2,c} - \Delta\tilde{\omega}$. The heights of the two asteroids from the equatorial plane of the sun are

$$z_1 = r_c \sin(f_{2,c} - \Delta\tilde{\omega} + \omega_1) i_1,$$
$$z_2 = r_c \sin(f_{2,c} + \omega_2) i_2, \tag{4.11}$$

where r_c is the orbital radius at the orbital crossing given by

$$r_c = \frac{p_2}{1 + e_2 \cos f_{2,c}}. \tag{4.12}$$

Note that we still assume i_1 and i_2 are small, which leads to $\sin i_1 \approx i_1$, $\cos i_1 \approx 0$, $\sin i_2 \approx i_2$, and $\cos i_2 \approx 0$. The collision occurs if $|z_1 - z_2|$ is smaller than $(D_1 + D_2)/2$, where D_1 and D_2 are the diameters of the asteroid 1 and 2, respectively. Here, to increase the number of samples, we use the more loose condition of

$$|z_1 - z_2| < 5(D_1 + D_2). \tag{4.13}$$

If Eqs. (4.10) and (4.13) are satisfied, we calculate the relative velocity between the two asteroids v_{12} at the orbital crossing as follows:

$$v_{12}^2 = GM_s\Bigg[\left(\frac{e_1 \sin(f_{2,c} - \Delta\tilde{\omega})}{\sqrt{p_1}} - \frac{e_2 \sin f_{2,c}}{\sqrt{p_2}}\right)^2 + \left(\frac{\sqrt{p_1}}{r_c} - \frac{\sqrt{p_2}}{r_c}\right)^2$$
$$+ \left(\frac{i_1[\cos(f_{2,c} - \Delta\tilde{\omega} + \omega_1) + e_1 \cos\omega_1]}{\sqrt{p_1}} - \frac{i_2[\cos(f_{2,c} + \omega_2) + e_2 \cos\omega_2]}{\sqrt{p_2}}\right)^2\Bigg]. \tag{4.14}$$

We evaluate whether the orbital crossing occurs or not using Eqs. (4.10) and (4.13) for each pair of all observed asteroids in the main belt, and we calculate possible impact velocities as Eq. (4.14) if Eqs. (4.10) and (4.13) are satisfied. The orbital elements of asteroids are obtained from JPL Small-Body Database. Figure 4.1 shows the distribution of the possible impact velocities between main-belt asteroids in the present main belt. The average impact velocity is calculated as 5 km/s, which is consistent with the rough estimation using the average eccentricity and inclination. Fraction of impact velocities smaller than 400 m/s is only about 0.2%. The impact simulations in Sect. 3.3 show that equal-mass and non-destructive impacts require $v_{\rm imp} < 400$ m/s, and thus similar-mass impacts in the present environment mainly result in catastrophic destruction.

4.1.3 Shapes Formed in the Primordial and Present Environments

The impact velocity is comparable to the escape velocity from asteroids with the diameter $D = 100$ km in the primordial environment before the formation of Mars-mass protoplanets (see Sect. 4.1.2). Asteroids with $D > 100$ km are mainly the largest bodies among asteroids (see Fig. 1.2), and thus they mainly collide with similar-mass or smaller asteroids. As we investigated in Sects. 3.3 and 3.4.1, similar-mass impacts, especially equal-mass impacts, produce various (bilobed, spherical, flat, elongated, and hemispherical) shapes including extremely elongated and flat shapes. High-mass-ratio impacts in the primordial environment merely result in crater formation or coalescence, and such impacts do not deform target asteroids. Although high-mass-ratio impacts are more frequent than similar-mass impacts, shapes of 100 km asteroids are probably determined by the last similar-mass impacts.

In the present environment, the typical impact velocity is much larger than the escape velocity from 100 km asteroids. Impacts with the impact velocity comparable to the escape velocity may occur, but the fraction of such impacts is about 0.2%; such impacts rarely occur (see Fig. 4.1). Relatively similar-mass impacts with the impact velocity much larger than the escape velocity result in destructive impacts. As we investigated in Sect. 3.5, destructive impacts mainly produce spherical and bilobed shapes. Larger mass-ratio impacts may result in non-destructive impacts with overall deformation of target asteroids. However, as we investigated in Sect. 3.4.2, such impacts do not produce irregular shapes of target asteroids with $c/a \leq 0.6$.

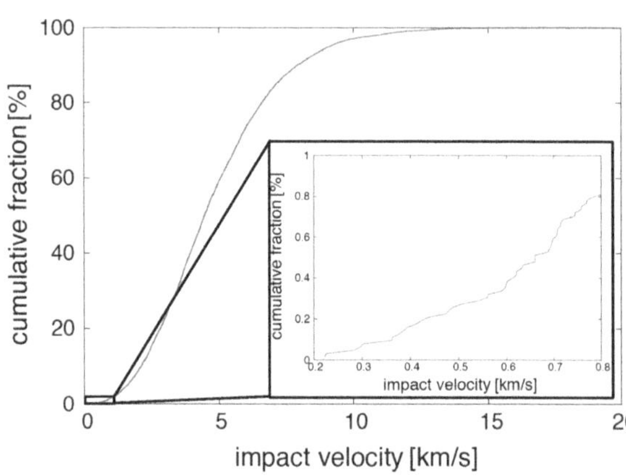

Fig. 4.1 Cumulative distribution of impact velocities of possible collisions between main-belt asteroids. The horizontal axis represents estimated impact velocity, and the vertical axis represents fraction of impact velocities that are smaller than the value in the horizontal axis. Blow-up figure shows the distribution for lower impact velocities

4.2 Comparison with Actual Asteroids

4.2.1 Databases Utilized in Our Analysis

4.2.1.1 DAMIT

Database of Asteroid Models from Inversion Technique[2] (DAMIT: Ďurech et al. 2010) is a database that mainly contains shape models of asteroids. The shape models of asteroids stored in DAMIT are mainly obtained from light curves of asteroids through light curve inversion technique (Kaasalainen and Torppa 2001; Kaasalainen et al. 2001). As of November 2018, DAMIT provides the shape models of 1,609 asteroids.

Light curve inversion technique accurately measures shapes of asteroids, at least the axis ratio b/a of elongated asteroids. For example, b/a of the asteroid Itokawa obtained from light curves is about 0.5 (Kaasalainen et al. 2003), while that measured by in-situ observation is 0.55 (Fujiwara et al. 2006). Thus the shape models obtained from light curves have enough accuracy to analyze the axis ratios of asteroids.

The shape models in DAMIT are represented as polyhedrons with triangular surface facets. For each asteroid, DAMIT provides a file that contains the information of coordinates x, y, and z on surface of a shape model of the asteroid. We utilize the coordinates of the surfaces of the shape models to derive the axis ratios of asteroids. To measure the axis ratios of the shape models of asteroids, we follow the bottom-up method (Fujiwara et al. 1978, see also Sect. 3.2). Note that the coordinates x, y, and z for the most shape models are normalized by volume of each shape model, so that we cannot achieve the information of sizes of asteroids from DAMIT.

Figure 4.2 shows the orbital semi-major axis and eccentricity of asteroids of which shape models are stored in DAMIT. From Fig. 4.2, we notice that the most asteroids in DAMIT are the main-belt asteroids, but some of them are the near-Earth asteroids, Hilda asteroids, and Trojan asteroids.

Figure 4.3 shows the ratio of the number of the asteroids that are included in DAMIT to that of all observed asteroids. We notice that DAMIT contains more than 50% of asteroids with diameters D larger than 100 km, and thus shapes of the almost all of the asteroids $\geq$100 km are measured. However, shapes of only about 10% of asteroids with $D > 10$ km are measured. For asteroids $\leq$10 km, the fraction of asteroids with known shapes is less than 1%.

4.2.1.2 JPL Small-Body Database

Small-Body Database of Jet Propulsion Laboratory (JPL) provides orbital and physical information of asteroids. The physical information contains diameter, absolute magnitude, albedo, and rotation period. We utilize the diameters of asteroids listed

[2]http://astro.troja.mff.cuni.cz/projects/asteroids3D/web.php.

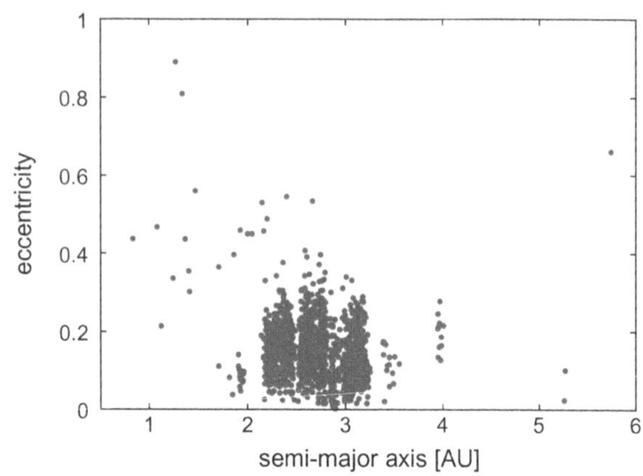

Fig. 4.2 Orbital elements of asteroids of which shape models are stored in DAMIT. The horizontal axis represents the orbital semi-major axis of the asteroids, and the vertical axis represents the orbital eccentricity of the asteroids

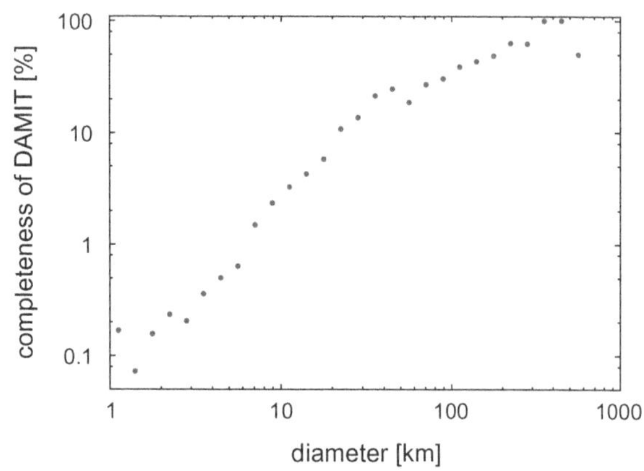

Fig. 4.3 Ratio of the number of DAMIT asteroids to that of all observed asteroids. The vertical axis represents the fraction, and the horizontal axis represents the diameter of asteroids

in JPL Small-Body Database to distinguish shapes of larger asteroids and smaller asteroids.

Diameters of asteroids cannot be determined solely by absolute magnitude in visible light, because the absolute magnitude depends on both of diameter and albedo and albedo is not well determined. Thus, the almost all of the diameters of asteroids in JPL Small-Body Database are obtained from NEOWISE mission. NEOWISE mission was conducted by the spacecraft Wide-field Infrared Survey Explorer (WISE). WISE observed various objects including asteroids in infrared wavelengths. Magnitude of asteroids in infrared wavelength is mainly determined by temperature of asteroids and not strongly depends on albedo. Therefore, if we utilize both the magnitudes in visible and infrared wavelength, we can accurately determine diameters of asteroids.

4.2.1.3 AstDyS-2

Asteroid Dynamics Site-2[3] (AstDyS-2) summarizes orbital information of asteroids. The information in AstDyS-2 includes that of asteroid families. Using this database, we can know that each asteroid is belonging to any family or not and name of a family that each asteroid belongs to. Asteroid families were formed through relatively recent impacts, and thus we distinguish asteroids that were affected by recent impacts and those are not affected using the family information obtained from AstDyS-2.

Family classification of AstDyS-2 is based on Milani et al. (2014). Note that results of family classification depend on classification method. Throughout this study, we utilize the family information obtained from AstDyS-2.

4.2.2 Fraction of Family Asteroids

Asteroids in asteroid families were formed in recent impacts. Although asteroid families are dissipated with typical timescale of about 1 Gyr, fraction of family asteroids may tell us how many asteroids were affected by recent destructive impacts. Thus, we calculate the fraction of family asteroids for various diameters of asteroids. We use the information of asteroid families obtained from AstDyS-2 and the diameters of asteroids obtained from JPL Small Body Database.

Shapes of asteroids in asteroid families were mainly formed through recent destructive impacts. However, the largest bodies of cratering families are not the case because those shapes were not affected by the cratering-family-forming impacts. Here, we consider families with $M_{\rm lr}/M_{\rm pb} > 0.8$ as cratering families, where $M_{\rm lr}$ is the mass of the largest bodies among asteroids in each family and $M_{\rm pb}$ is the total mass of each family, i.e., the mass of a parent body. We calculate the ratio $M_{\rm lr}/M_{\rm pb}$ using diameters of member asteroids as follows:

$$\frac{M_{\rm lr}}{M_{\rm pb}} = \frac{D_{\rm lr}^3}{\sum_n D_n^3}, \tag{4.15}$$

where $D_{\rm lr}$ is diameters of the largest bodies, D_n is diameters of member asteroids, and $\sum_n$ denotes the summation for all member asteroids. Here, we assume the density for all asteroids is the same. Hereafter, family asteroids mean asteroids in any families except for the largest asteroids of cratering families, and non-family asteroids mean the largest asteroids of cratering families in addition to asteroids that do not belong to any families.

Figure 4.4 shows dependence of the fraction of family asteroids on diameters of asteroids. As we notice from Fig. 4.4, the fraction is less than 10% for asteroids with $D > 100$ km, which is significantly smaller than that for smaller asteroids $\approx 30\%$. This is because the larger asteroids are difficult to be disrupted and collisions

[3] https://newton.spacedys.com/astdys2/index.php?pc=0.

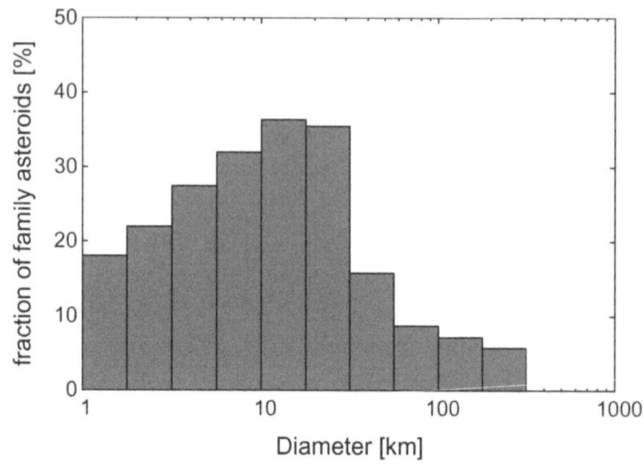

Fig. 4.4 Fraction of family asteroids. The vertical axis represents the fraction, i.e., the ratio of the number of family asteroids to that of all asteroids, and the horizontal axis represents diameters of asteroids

that destroy the larger asteroids are not frequent. Thus about 90% of asteroids with $D > 100$ km are probably not affected by recent destructive impacts. This is consistent with the collisional lifetime estimated in O'Brien and Greenberg (2005); the collisional lifetimes for asteroids >100 km are estimated to be about ten times longer than the age of the solar system, which means that only about one tenth of asteroids are destructed within the timescale comparable to the age of the solar system. Therefore, shapes of the larger asteroids may keep those formed in the primordial environment.

We also notice from Fig. 4.4 that the fraction decreases with decreasing diameters for asteroids with $D < 10$ km. The timescale of the Yarkovsky effect for the asteroids with $D = 1$ km is about ten times longer than the collisional lifetime (e.g., Morbidelli and Vokrouhlický 2003). Thus this tendency is probably not caused by the dissipation of families due to the Yarkovsky effect. Although we have not yet specified the reason of this tendency, we think that this tendency is merely caused by the error of orbital elements or incompleteness of family classification due to smaller sizes of asteroids.

4.2.3 Shapes of Asteroids Larger Than 100 km

As we showed in Sect. 4.2.2, the fraction of family asteroids with diameters $D \geq 100$ km is significantly smaller than that with $D < 100$ km, which means that shapes of the larger asteroids may keep those formed in the primordial environment. We showed in Sect. 4.1.2 that low-velocity impacts in the primordial environment mainly result in non-destructive impacts, while high-velocity impacts in the present environment mainly result in destructive impacts. Thus, the comparison between shapes of non-family and family asteroids with $D \geq 100$ km may suggest the difference of the formation era, which leads to constrain when these asteroids were formed. We also discuss the comparison between the results of our collisional simulations and shapes of actual asteroids.

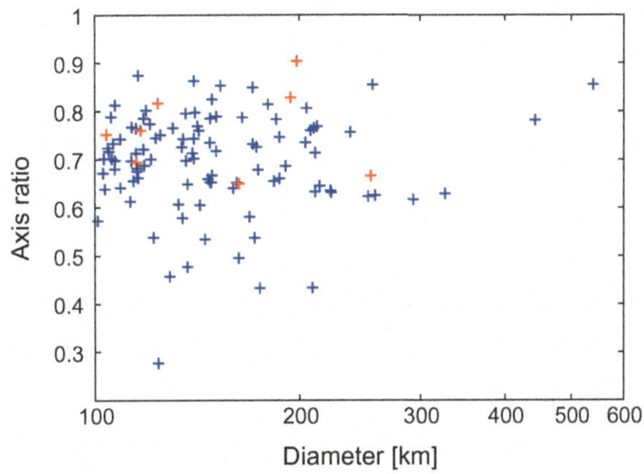

Fig. 4.5 Axis ratio c/a of asteroids with diameters D larger than 100 km. The vertical axis represents c/a, and the horizontal axis represents the diameter. The blue points show non-family asteroids, while the red points show family asteroids

Figure 4.5 shows the axis ratio c/a of non-family and family asteroids with diameters >100 km. All family asteroids, which were affected by recent destructive impacts, have the axis ratio $c/a > 0.6$, i.e., they have rounded shapes, while some of non-family asteroids have irregular shapes with $c/a < 0.6$. This may suggest the difference of the formation environment. However, this tendency is not statistically significant. The number of family asteroids that have irregular shapes with $c/a < 0.6$ is zero among eight family asteroids, while the number of non-family asteroids that have irregular shapes with $c/a < 0.6$ is 12 among 107 non-family asteroids. The fraction of the irregular shapes of the family asteroids is not so different from that of the non-family asteroids.

Figure 4.6a and b show the axis ratios of actual non-family and family asteroids, respectively. Figure 4.6b again shows that all of shapes of the family asteroids with $D > 100$ km are rounded shapes with $c/a > 0.6$, while Fig. 4.6a shows that shapes of the non-family asteroids include an extremely elongated shape with $b/a \approx 0.4$ or extremely flat shapes with $c/a \approx 0.4$. From the comparison between shapes of the remnants produced through our simulations of the destructive impacts (Fig. 4.6d) and those of actual non-family asteroids (Fig. 4.6a), we notice that the destructive impacts, especially the unequal-mass impacts, are difficult to produce the flat shapes of the actual asteroids with $c/a \approx 0.4$. Note that the impacts with $q = 1$ in Fig. 4.6d has the impact velocity of $v_{\rm imp} = 350$ m/s, which is still difficult to occur in the present environment. In contrast, Fig. 4.6c shows that the non-destructive and equal-mass impacts can produce flat shapes with $c/a \approx 0.4$.

As we discussed in Sect. 4.1.2, collisions in the present environment mainly result in destructive impacts, which do not lead to the formation of flat shapes. In contrast, collisions in the primordial environment mainly result in non-destructive impacts, so that similar-mass impacts in the primordial environment lead to the formation of flat shapes. Therefore, the flat shapes of the non-family asteroids with $D > 100$ km are likely to be formed in the primordial environment before the formation of Jupiter, and remain the same until today.

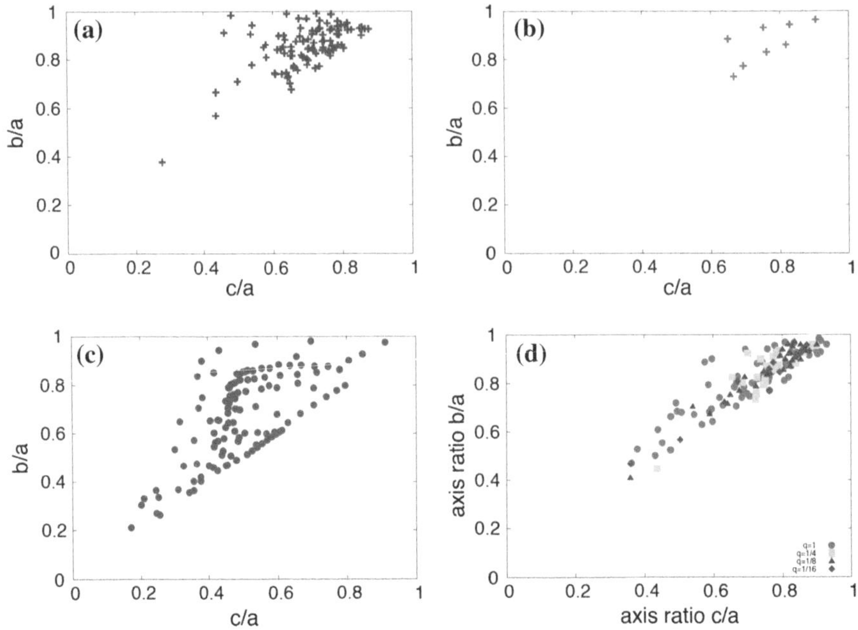

Fig. 4.6 Shape distributions of actual asteroids and the remnants produced through our simulations. In each panel, the horizontal axis represents the axis ratio c/a, and the vertical axis represents the axis ratio b/a. The panel **a** shows the shape distribution of actual non-family asteroids with $D > 100$ km, and the panel **b** shows that of actual family asteroids. The panels **c** and **d** show those of the remnants produced through our simulations of the non-destructive and equal-mass impacts (see Fig. 3.15) and the destructive impacts (see Fig. 3.24), respectively

Family asteroids were formed through recent destructive impacts. However, non-family asteroids were not necessarily formed in the primordial environment. They may have been formed through recent destructive impacts, but they are no longer family asteroids due to, for example, scattering through mean motion resonances with gas giants. The information of shapes, especially flat shapes of asteroids, tells us they were surely formed in the primordial environment through low-velocity impacts.

4.2.4 *Shapes of Asteroids Larger Than 10 km*

The fraction of family asteroids with $D > 10$ km is much higher than that of asteroids with $D > 100$ km (see Fig. 4.4). Thus, it is less likely that shapes of the asteroids with $D > 10$ km also keep those formed in the primordial environment. However, O'Brien and Greenberg (2005) show that the collisional lifetimes for the asteroids with $D = 10$ km are comparable to the age of the solar system, which suggests that about half of such asteroids may not experience recent destructive impacts. Moreover,

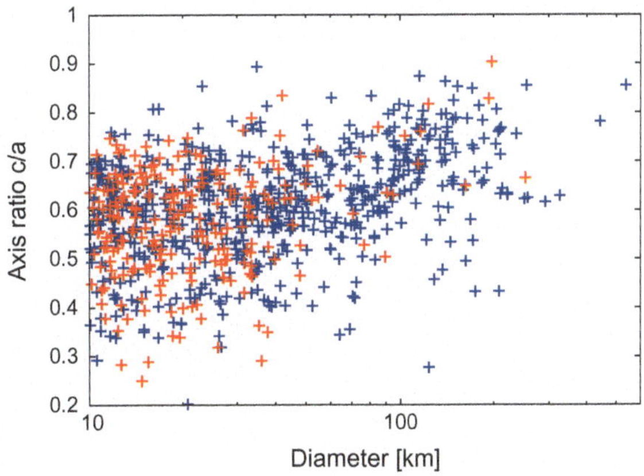

Fig. 4.7 Axis ratio c/a of asteroids with $D > 10$ km. The vertical axis represents c/a, and the horizontal axis represents the diameter. The blue points show non-family asteroids, while the red points show family asteroids

Fig. 4.4 shows that about two third of the asteroids with $D \approx 10$ km still do not belong to any family, which means that they may not be affected by recent impacts. Thus, we may also find the difference of shapes between family and non-family asteroids with $D > 10$ km.

Figure 4.7 is the same as Fig. 4.5, but for the asteroids with $D > 10$ km. As we notice from Fig. 4.7, the difference of c/a between the family and non-family asteroids is not apparent compared to the asteroids with $D > 100$ km. The fraction of the asteroids with $c/a < 0.6$ is similar for the family and non-family asteroids. The number of the family asteroids with $c/a < 0.6$ is 120 among 237 family asteroids; the fraction is 50.6%. The number of the non-family asteroids with $c/a < 0.6$ is 255 among 591 non-family asteroids; the fraction is 43.1%. Non-destructive impacts produce various irregular shapes (see Fig. 3.15) and destructive impacts produce bilobed shapes (see Fig. 3.24); both non-destructive and destructive impacts produce irregular shapes with smaller c/a. This suggests that we cannot distinguish family and non-family asteroids only with the ratio c/a.

As we investigated in Sect. 3.5, destructive impacts are difficult to produce flat shapes. Thus the fraction of flat shapes of family asteroids may be different from that of non-family asteroids. Figure 4.8 shows the shape distribution of the asteroids with $D > 10$ km. Here, we define flat shapes as the shape with $c/a < 0.6$ and $b/a > 0.8$, and count the number of the flat shapes for the non-family and family asteroids. We find 50 non-family asteroids with the flat shapes among 591 non-family asteroids; the fraction of the flat shapes is 8.5%. In contrast, we find 9 family asteroids with the flat shapes among 237 family asteroids; the fraction is 3.8%. Therefore, the fraction of the flat shapes of the non-family asteroids is more than twice as high as that of the family asteroids; the fraction of the flat shapes of the family asteroids, which are affected by recent destructive impacts, is significantly small. It is consistence with the result of our simulations showing that destructive impacts are difficult to produce flat shapes.

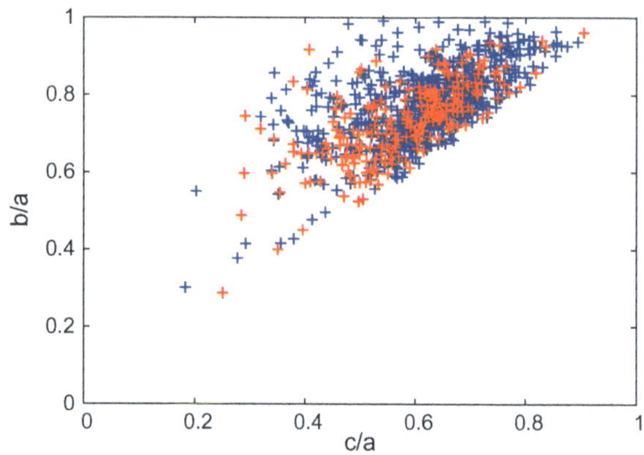

Fig. 4.8 Shape distribution of asteroids with $D > 10$ km. The horizontal axis represents the ratio c/a, and the vertical axis represents the ratio b/a. The blue points show non-family asteroids, while red points show family asteroids

Our simulations predict that destructive impacts do not produce flat shapes, and thus we expect that the family asteroids do not include flat shapes. However, Fig. 4.8 shows that there are several family asteroids with flat shapes. This may be caused by observational errors; it might be difficult to distinguish flat shapes and elongated shapes solely by light curves because both flat and elongated shapes cause brightness variations due to rotations, and flat shapes of the family asteroids may be elongated shapes.

Although we have not yet specified the reason why the family asteroids with $D > 10$ km include flat shapes, there is one possible mechanism. Dispersion speeds of family asteroids just after shattering of parent bodies are estimated to be ~100 m/s (e.g., Michel et al. 2002); the dispersion speeds are comparable to impact velocities of non-destructive impacts in the primordial environment. Shuttered bodies return to the place where family-forming impacts occur after one or more revolutions around the sun, and they may collide with each other with low impact velocities, which may result in the formation of flat shapes of family asteroids if colliding two bodies have similar masses.

The reason why the family asteroids with $D > 100$ km do not include flat shapes is expected to be lack of similar-mass family members. The family asteroids with $D > 100$ km are mainly the largest bodies among family members. The number of family members increases with decreasing sizes of members, and thus the number of similar-mass members for $D > 100$ km is small: Similar-mass impacts between family members with $D > 100$ km are difficult to occur, which may lead to the lack of flat family asteroids. We counted the number of similar-mass asteroids (the mass ratio is larger than 0.5) among members in the same families for eight family asteroids with $D > 100$ km ((5) Astraea, (15) Eunomia, (423) Diotima, (85) Io, (104) Klymene, (24) Themis, (171) Ophelia, and (221) Eos). The number of similar-mass members is as follows: 1 for (5) Astraea, 0 for (15) Eunomia, 0 for (423) Diotima, 1 for (85) Io, 3 for (104) Klymene, 0 for (24) Themis, 4 for (171) Ophelia, and 5 for (221) Eos. Thus the number of similar-mass family members is in fact small for the family asteroids with $D > 100$ km.

In contrast, the number of similar-mass members for family asteroids with $D > 10$ km is relatively large because they are probably not the largest bodies among asteroids in the same families. We also counted the number of similar-mass members for nine family asteroids with the flat shapes and $D > 10$ km ((135) Hertha, (355) Gabriella, (1040) Klumpkea, (1286) Banachiewicza, (2962) Otto, (6905) Miyazaki, (3261) Tvardovskij, (608) Adlfine, and (1836) Komarov). Six asteroids among them indeed have large number of similar-mass members: 133 for (1286) Banachiewicza, 9 for (2962) Otto, 62 for (6905) Miyazaki, 90 for (3261) Tvardovskij, 151 for (608) Adlfine, and 8 for (1836) Komarov. Thus the flat shapes of these family asteroids are possible to be formed through low-velocity and similar-mass impacts between member asteroids. However, the number of similar-mass members for three of them is very small: 0 for (135) Hertha, 0 for (355) Gabriella, and 3 for (1040) Klumpkea. We do not know any mechanism to produce the flat shapes of these family asteroids, and they may be interlopers that were non-family asteroids in past.

4.2.5 Direction of Rotation Axis

We investigated the collisional formation of elongated and flat shapes through the equal-mass and low-velocity impacts in Sect. 3.3. Our simulations predict that the rotation axis of elongated bodies are perpendicular to the major axis and those of flat bodies are perpendicular to the minor axis if these bodies are formed through equal-mass and low-velocity impacts. To verify our prediction, we investigate angles between rotation axis and the major or minor axis of actual asteroids.

Figure 4.9a shows the angles between the rotation axis and the major axis of elongated asteroids with $b/a < 0.6$ and $c/a < 0.6$. From Fig. 4.9a, we notice that

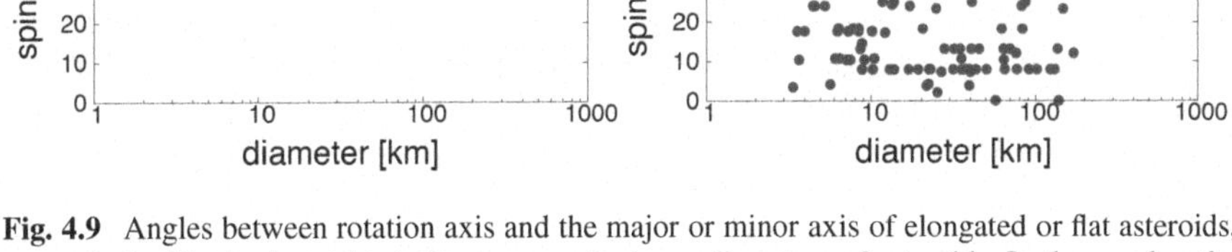

Fig. 4.9 Angles between rotation axis and the major or minor axis of elongated or flat asteroids, respectively. The horizontal axis for two panels shows diameters of asteroids. In the panel **a**, the vertical axis represents the angles between the rotation axis and the major axis of elongated asteroids with $b/a < 0.6$ and $c/a < 0.6$. In the panel **b**, the vertical axis represents the angles between the rotation axis and the minor axis of flat asteroids with $b/a > 0.8$ and $c/a < 0.6$

all the elongated asteroids have the rotation axis almost perpendicular to the major axis, which means that the rotation axis almost aligns with the minor axis. The directions of the rotation axis of the elongated asteroids are consistent with those of the elongated remnants formed through our simulations of the equal-mass and low-velocity impacts (see e.g., Fig. 3.2). From Fig. 4.9b, we notice that the almost all of the flat asteroids have the rotation axis that is parallel to the minor axis. The directions of the rotation axis of the flat remnants produced through our simulations of the equal-mass and low-velocity impacts are perpendicular to the minor axis (see Fig. 3.8), and thus the directions of the rotation axis of the actual flat asteroids are inconsistent with the results of our simulations.

In Sect. 4.2.3, we argued that the flat asteroids with $D > 100$ km were formed in the primordial environment before the formation of Jupiter through similar-mass and low-velocity impacts (see Fig. 3.8). The rotation axis of the produced flat remnants is perpendicular to the minor axis and they rotate around the major axis. Rotation energy of the rotation around the major axis is larger than that around the minor axis, and thus internal energy dissipation causes gradual alignment of the rotation axis with the minor axis. Internal energy dissipation is mainly caused by deformation stresses due to gyroscopic torque and stresses due to deformation of centrifugal bulges during wobble motions (Burns and Safronov 1973). Note that the flat remnant in Fig. 3.8 also does the wobble motion with much longer period than the duration of the simulation. Burns and Safronov (1973) estimated timescale of the alignment as

$$\tau_{\mathrm{align}} = \frac{\mu Q}{\rho K_3^2 R_{\mathrm{ast}}^2 \omega_{\mathrm{rot}}^3}, \tag{4.16}$$

where μ is the rigidity of asteroids, ρ is its density, R_{ast} is its radius, and ω_{rot} is its rotational angular frequency. Q is a material's quality factor, which is roughly defined as the ratio of the total stored strain energy to the energy loss per cycle of the wobble motion. K_3^2 is a shape factor, and $K_3^2 \approx 10^{-2}$ for spherical bodies and $K_3^2 \approx 0.1[(a-c)/c]^2$ for irregular shaped bodies. With probable parameters shown in Burns and Safronov (1973), the alignment timescale is estimated to be

$$\tau_{\mathrm{align}} = 1.0 \times 10^6 \left(\frac{R_{\mathrm{ast}}}{50\,\mathrm{km}}\right)^{-2} \left(\frac{T_{\mathrm{rot}}}{8\,\mathrm{hour}}\right)^3 \mathrm{year}, \tag{4.17}$$

where T_{rot} is rotation periods. Thus, the timescale for the alignment of the rotation axis with the minor axis is about 1 Myr, which is much shorter than the age of the solar system. Therefore, we suggest that the flat asteroids with $D > 100$ km were formed in the primordial environment and initially had the rotation axis perpendicular to the minor axis, but the rotation axis now aligns with the minor axis due to the internal energy dissipation.

References

J.A. Burns, V.S. Safronov, Asteroid nutation angles. Mon. Not. R. Astron. Soc. **165**, 403 (1973). https://doi.org/10.1093/mnras/165.4.403

J. Ďurech, V. Sidorin, M. Kaasalainen, DAMIT: a database of asteroid models. Astron. Astrophys. **513**, A46 (2010). https://doi.org/10.1051/0004-6361/200912693

A. Fujiwara, G. Kamimoto, A. Tsukamoto, Expected shape distribution of asteroids obtained from laboratory impact experiments. Nature **272**, 602 (1978). https://doi.org/10.1038/272602a0

A. Fujiwara, J. Kawaguchi, D.K. Yeomans, M. Abe, T. Mukai, T. Okada, J. Saito, H. Yano, M. Yoshikawa, D.J. Scheeres, O. Barnouin-Jha, A.F. Cheng, H. Demura, R.W. Gaskell, N. Hirata, H. Ikeda, T. Kominato, H. Miyamoto, A.M. Nakamura, R. Nakamura, S. Sasaki, K. Uesugi, The rubble-pile asteroid Itokawa as observed by Hayabusa. Science **312**, 1330–1334 (2006). https://doi.org/10.1126/science.1125841

C. Hayashi, Structure of the solar nebula, growth and decay of magnetic fields and effects of magnetic and turbulent viscosities on the nebula. Prog. Theor. Phys. Suppl. **70**, 35–53 (1981). https://doi.org/10.1143/PTPS.70.35

S. Ida, J. Makino, Scattering of planetesimals by a protoplanet—slowing down of runaway growth. Icarus **106**, 210 (1993). https://doi.org/10.1006/icar.1993.1167

M. Kaasalainen, J. Torppa, Optimization methods for asteroid lightcurve inversion. I. Shape Determination. Icarus **153**, 24–36 (2001). https://doi.org/10.1006/icar.2001.6673

M. Kaasalainen, J. Torppa, K. Muinonen, Optimization methods for asteroid lightcurve inversion. II. Complet. Inverse Problem. Icarus **153**, 37–51 (2001). https://doi.org/10.1006/icar.2001.6674

M. Kaasalainen, T. Kwiatkowski, M. Abe, J. Piironen, T. Nakamura, Y. Ohba, B. Dermawan, T. Farnham, F. Colas, S. Lowry, P. Weissman, R.J. Whiteley, D.J. Tholen, S.M. Larson, M. Yoshikawa, I. Toth, F.P. Velichko, CCD photometry and model of MUSES-C target (25143) 1998 SF36. Astron. Astrophys. **405**, L29–L32 (2003). https://doi.org/10.1051/0004-6361:20030819

H. Kobayashi, S. Ida, The effects of a stellar encounter on a planetesimal disk. Icarus **153**, 416–429 (2001). https://doi.org/10.1006/icar.2001.6700

E. Kokubo, S. Ida, Formation of protoplanet systems and diversity of planetary systems. Astrophys. J. **581**, 666–680 (2002). https://doi.org/10.1086/344105

P. Michel, W. Benz, P. Tanga, D.C. Richardson, Collisions and gravitational reaccumulation: forming asteroid families and satellites. Science **294**, 1696–1700 (2001). https://doi.org/10.1126/science.1065189

A. Milani, A. Cellino, Z. Knežević, B. Novaković, F. Spoto, P. Paolicchi, Asteroid families classification: exploiting very large datasets. Icarus **239**, 46–73 (2014). https://doi.org/10.1016/j.icarus.2014.05.039

A. Morbidelli, D. Vokrouhlický, The Yarkovsky-driven origin of near-earth asteroids. Icarus **163**, 120–134 (2003). https://doi.org/10.1016/S0019-1035(03)00047-2

D.P. O'Brien, R. Greenberg, The collisional and dynamical evolution of the main-belt and NEA size distributions. Icarus **178**, 179–212 (2005). https://doi.org/10.1016/j.icarus.2005.04.001

T. Ueda, H. Kobayashi, T. Takeuchi, D. Ishihara, T. Kondo, H. Kaneda, Size dependence of dust distribution around the earth orbit. Astron. J. **153**, 232 (2017). https://doi.org/10.3847/1538-3881/aa5ff3

S.J. Weidenschilling, Aerodynamics of solid bodies in the solar nebula. Mon. Not. R. Astron. Soc. **180**, 57–70 (1977). https://doi.org/10.1093/mnras/180.1.57

F.L. Whipple. On certain aerodynamic processes for asteroids and comets, in *From Plasma to Planet*, ed. by A. Elvius (1972), p. 211

D.P. Whitmire, J.J. Matese, L. Criswell, S. Mikkola, Habitable planet formation in binary star systems. Icarus **132**, 196–203 (1998). https://doi.org/10.1006/icar.1998.5900

Chapter 5
Application: Extremely Elongated Shape Of 1I/'Oumuamua

Abstract We showed that similar-mass and non-destructive impacts can produce extremely elongated shapes of asteroids. Recently an interstellar object 1I/'Oumuamua was discovered, and its light curve shows that 1I/'Oumuamua has the size of about 100 m and an extremely elongated shape with the ratio of intermediate to major axis lengths less than 0.3. In this chapter, we introduce results of our simulations that investigate detailed impact conditions to form such extremely elongated object. The results show that the collisional formation of the extremely elongated shape requires impact velocities less than 40 cm/s, impact angles less than 30 degrees, ratios of impactor mass to target mass larger than 0.5, and friction angles of granular material composing impacting objects larger than 40 degrees. Such low impact velocities are realized in extremely young protoplanetary disks without planetesimals larger than 7 km. Thus, we suggest that 1I/'Oumuamua came to our solar system from such an extremely young protoplanetary disk.

Keywords 1I/'Oumuamua · Interstellar object · Extremely elongated shape · Formation environment

5.1 Introduction of 1I/'Oumuamua

The telescope Pan-STARRS1 at the Haleakala Observatory, Hawaii, discovered 1I/'Oumuamua on October 2017 (Meech et al. 2017). 1I/'Oumuamua has high orbital eccentricity of $e \approx 1.2$, which suggests that 1I/'Oumuamua came from interstellar space. During era of planet formation, giant planets eject large amounts of planetesimals from planetary systems (Bottke et al. 2005; Raymond et al. 2018; Jackson et al. 2018). Moreover, speed of 1I/'Oumuamua relative to the sun at infinity is calculated to be $v_\infty \approx 26\,\mathrm{km/s}$, and it is close to random speeds of solar neighborhoods (Meech et al. 2017). Therefore, we expect that 1I/'Oumuamua is one of planetesimals ejected from a nearby planetary system (Meech et al. 2017; de la Fuente Marcos and de la Fuente Marcos 2017).

1I/'Oumuamua has absolute magnitude of $H_\mathrm{V} \approx 22$ mag. If we assume that albedo of the object is 0.04, $H_\mathrm{V} \approx 22$ mag corresponds to mean radius of ≈ 100 m (Meech

K. Sugiura, *Development of a Numerical Simulation Method for Rocky Body Impacts and Theoretical Analysis of Asteroidal Shapes*, Springer Theses,
https://doi.org/10.1007/978-981-15-3722-6_5

et al. 2017; Bolin et al. 2018; Bannister et al. 2017). Light curve observations show that 1I/'Oumuamua rotates with rotation period of about 8 h (e.g., Bolin et al. 2018). Spectral features or slopes of this object agree with those of D-type asteroids or comets (Meech et al. 2017). In spite of facts that no imaging observations confirm cometary activity of 1I/'Oumuamua (Meech et al. 2017; Ye et al. 2017; Jewitt et al. 2017), detailed measurements of its orbit show that the orbit slightly deviates from that predicted only by the gravity of the sun and requires non-gravitational acceleration (Micheli et al. 2018), which suggests ejection of mass from its surface. Thus whether its composition is asteroid- or comet-lke is still under debate.

Interestingly, 1I/'Oumuamua shows very large amplitude of its light curve $\approx$ 2.5 mag, which is much larger than those of any asteroids in the solar system (Meech et al. 2017). Luminosity of small bodies mainly changes due to change of those apparent cross sections, which are caused by rotations. If small bodies rotate around those minor axes, light curve amplitudes correspond to the ratios of intermediate axis lengths b to major axis lengths a. Therefore, the large amplitude of the light curve of 1I/'Oumuamua suggests that the object has an extremely elongated shape. Although estimation of b/a shows variations among several studies because of short observational period and observational errors, the axis ratio b/a is estimated as follows: $b/a < 0.19$ (Bannister et al. 2017), $0.10 < b/a < 0.29$ (Bolin et al. 2018), $b/a < 0.22$ (Drahus et al. 2018), $b/a < 0.20$ (Fraser et al. 2018), $b/a < 0.33$ (Knight et al. 2017), and $b/a \approx 0.1$ (Meech et al. 2017). Therefore, 1I/'Oumuamua probably has an extremely elongated shape with the ratio $b/a < 0.3$.

We find several mechanisms explaining extremely elongated objects in previous studies. Impact destruction of a monolithic body may produce some small and monolithic fragments with $b/a < 0.3$ (Michikami et al. 2016). Tidal elongation through close encounters with gas giants or central stars also produce such extremely elongated objects (Walsh and Richardson 2006). Spinup of rubble pile bodies results in rapid rotation states and they may deform into elongated shapes (Richardson et al. 2005). Recently it is suggested that initially elongated objects evolve into more elongated shapes through surface excavation due to micro particle collisions, which results in the formation of extremely elongated objects (Domokos et al. 2009, 2017).

As shown in Sect. 3.3, low-velocity collisions between similar-mass bodies can produce elongated objects (see also e.g., Leinhardt et al. 2000, 2010; Sugiura et al. 2018). In this chapter, we investigate the possibility to form extremely elongated objects through collisions in more detail. 1I/'Oumuamua has the size of about 100 m, and thus we consider impacts between 100 m-sized objects. Then we constrain detailed impact conditions, i.e., the impact velocity, the impact angle, the ratio of masses of two impacting objects, and the friction angle, required for the formation of extremely elongated shapes with $b/a < 0.3$. Hereafter, we call remnants with such shapes extremely elongated remnants (EERs).

5.2 Initial Conditions of Simulations for 1I/‘Oumuamua

We use almost the same methods and initial conditions as in Chap. 3. Here, we describe different points from Chap. 3.

We mainly focus on merging collisions and investigate shapes of the largest remnants. Thus radius of target bodies is set to $R_t = 50\,\mathrm{m}$ so that sizes of the largest remnants become similar to that of 1I/‘Oumuamua. As in Sect. 3.3, we resolve a target body using about 50,000 SPH particles.

We assume that initial bodies have primordial property before they melt due to the decay energy of ^{26}Al, i.e, they do not have the tensile strength of solid body and shear strength is determined by the friction of granular material (Richardson et al. 2002). Thus, we initially set damage parameters D of all SPH particles to unity.

For impact simulations of meter-sized bodies, straightforward utilization of the Tillotson equation of state and the parameter set for basalt (Benz and Asphaug 1999) causes a problem. The sound speed of basaltic bodies C_s is about $3\,\mathrm{km/s}$ and a smoothing length h is about 1 m, and thus a time step determined by Eq. (2.98) is $\sim 10^{-3}$ s. Completion of simulations requires computation periods longer than the time scale of gravitational reaccumulation $\approx 1,600\,\mathrm{s}$ estimated in Eq. (3.1), which needs enormous number of time steps and computational cost. However, results of impacts between granular bodies with impact velocities much smaller than C_s do not depend on C_s, and we can artificially reduce C_s without changing results of simulations. Validity of the reduction of C_s is explained as follows: acceleration (or the equation of motion) of each SPH particle is calculated using the self-gravity, the pressure, and the deviatoric stress tensor. The pressure of bodies is mainly determined by the self-gravity for low-velocity impacts (see also Sect. 3.3.5). Since the deviatoric stress tensor described by the friction of granular material is proportional to the pressure, the deviatoric stress tensor is also determined by the self-gravity. Therefore, the self-gravity solely describes the equation of motion of each SPH particle, and the results of such impacts are mainly independent of the sound speed C_s.

We conduct impact simulations with the same conditions (the mass ratio of two impacting bodies $q = 1$, the impact angle $\theta_{\mathrm{imp}} = 15°$, the angle of friction $\phi_d = 40°$, the impact velocity $v_{\mathrm{imp}} = 20\,\mathrm{cm/s}$) but with different values of C_s. For the Tillotson equation of state, $C_s \approx \sqrt{A_{\mathrm{Til}}/\rho_0}$, where A_{Til} and ρ_0 are a Tillotson parameter and uncompressed density, respectively (see e.g., Eq. (2.99)). We change C_s by changing A_{Til}. Note that $B_{\mathrm{Til}} \approx A_{\mathrm{Til}}$, where B_{Til} is also a Tillotson parameter, and thus we also change value of B_{Til}. Figure 5.1a and b show that impact outcomes produced through the impact simulations with different values of A_{Til} and B_{Til} are very similar. Thus, we can reduce C_s, or A_{Til} and B_{Til}, without changing results of impact simulations.

Hereafter, we set A_{Til} and B_{Til} to 2.67×10^4 Pa, which results in the reduced sound speed of $C_s \approx 3\,\mathrm{m/s}$. For our simulations in this Chapter, we set impact velocities smaller than $40\,\mathrm{cm/s}$, and thus the sound speed $C_s \approx 3\,\mathrm{m/s}$ is still much larger than the impact velocities. The reduced sound speed allows us to conduct impact simulations with small granular bodies. Jutzi and Asphaug (2015) adopt the same approach using reduced sound speeds.

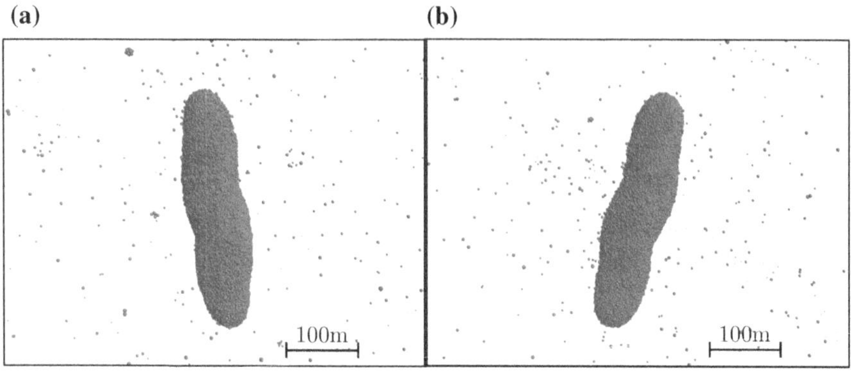

Fig. 5.1 Snapshots of the impact with $R_t = 50$ m, $q = 1$, $\phi_d = 40°$, $\theta_{\rm imp} = 15°$, and $v_{\rm imp} = 20$ cm/s at $t = 1.0 \times 10^5$ s. The panel **a** shows the result of the simulation with $A_{\rm Til} = B_{\rm Til} = 2.67 \times 10^4$ Pa, and the panel **b** shows that with $A_{\rm Til} = B_{\rm Til} = 2.67 \times 10^5$ Pa. Reprinted from Sugiura et al. (2019), Copyright (2019), with permission from Elsevier

5.3 Results of Simulations for 1I/'Oumuamua

We conduct various impact simulations with various mass ratios q, impact angles $\theta_{\rm imp}$, impact velocities $v_{\rm imp}$, and friction angles ϕ_d. Table 5.1 summarizes the conditions of the simulations.

5.3.1 Equal-Mass Impacts with $\phi_d = 40°$

Figure 5.2 shows snapshots of the impact simulation with $q = 1$, $\theta_{\rm imp} = 15°$, $v_{\rm imp} = 20$ cm/s, and $\phi_d = 40°$. The bodies are stretched through the collision (Fig. 5.2b, c), and the merging collision forms an EER (Fig. 5.2d, e). The EER has the axis ratio b/a

Table 5.1 Conditions of the simulations. $\Delta v_{\rm imp}$ and $\Delta\theta_{\rm imp}$ show increments of $v_{\rm imp}$ and $\theta_{\rm imp}$ in parameter surveys, respectively. Reprinted from Sugiura et al. (2019), Copyright (2019), with permission from Elsevier

q	ϕ_d (°)	$v_{\rm imp}$ (cm/s)	$\Delta v_{\rm imp}$ (cm/s)	$\theta_{\rm imp}$ (°)	$\Delta\theta_{\rm imp}$ (°)	Remark
1	40	12 – 36	3	5 – 30	5	Fig. 5.3a
1	40	15 – 30	1	7.5 – 20	2.5	Fig. 5.3b
1	50	12.5 – 40	2.5	5 – 40	5	Fig. 5.5
1	30	12.5 – 35	2.5	5 – 20	5	Fig. 5.6
0.5	50	15 – 35	2	10 – 35	5	Fig. 5.9
0.25	50	15 – 55	4	10 – 35	5	Fig. 5.10

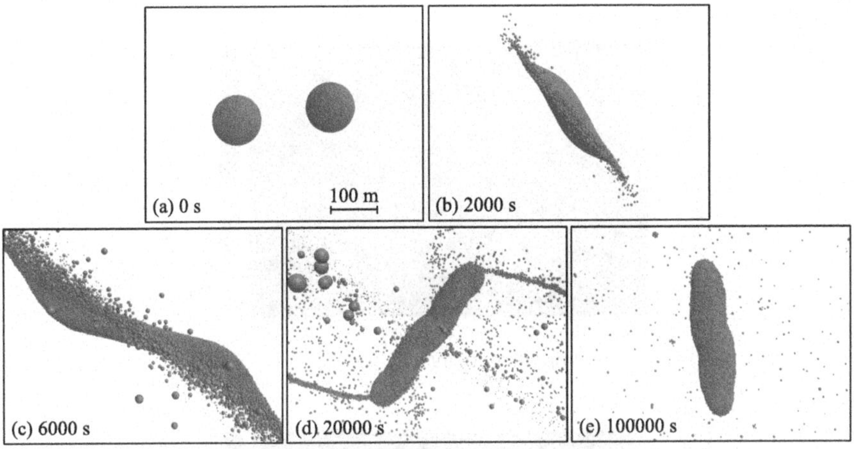

Fig. 5.2 Snapshots of the impact simulation with $q = 1$, $\theta_{imp} = 15°$, $v_{imp} = 20\,\mathrm{cm/s}$, and $\phi_d = 40°$. Reprinted from Sugiura et al. (2019), Copyright (2019), with permission from Elsevier

of 0.24 and the rotation period of 9.36 h. The axis ratio and the rotation period reproduce those of 1I/'Oumuamua. The extremely elongated shape seems to be unstable, but we confirm that the shape does not change until 1.0×10^5 s $\approx$ three rotations of the EER.

The largest remnant mainly becomes the most elongated remnant formed through each impact, and thus we investigate the axis ratios of the largest remnants. Figure 5.3 shows b/a of the largest remnants formed through various impacts with $q = 1$ and $\phi_d = 40°$. A rough parameter survey (Fig. 5.3a) shows that EERs are formed through impacts with $\theta_{imp} \approx 15°$ and $v_{imp} \approx 15 - 20\,\mathrm{cm/s}$. Figure 5.3b shows result of a more detailed parameter survey, and we find that EERs are formed through impacts with $v_{imp} \sin\theta_{imp} \approx 5.1$ cm/s. Here, all the largest remnants with $b/a < 0.3$ have the mass of $\approx 2.0 M_{target}$, which means almost complete merging.

According to the results in Sect. 3.3, we need large $v_{imp} \sin\theta_{imp}$ to form elongated bodies. However, impacts with too large $v_{imp} \sin\theta_{imp}$ result in fission of two impacting bodies, and they do not form EERs. $v_{imp} \sin\theta_{imp} \approx 5.1$ cm/s is sufficiently large to form extremely elongated objects and sufficiently small to form bodies with $\approx 2.0 M_{target}$. Even if $v_{imp} \sin\theta_{imp} \approx 5.1$ cm/s, catastrophic disruption mainly occurs for impacts with $v_{imp} \geq 25\,\mathrm{cm/s}$, and hit-and-run collisions mainly occur for impacts with $\theta_{imp} \geq 20°$; both kinds of impacts probably do not form elongated objects. Therefore, for impacts with $q = 1$ and $\phi_d = 40°$, we need the impact conditions of $\theta_{imp} \leq 20°$, $v_{imp} \leq 25\,\mathrm{m/s}$, and $v_{imp} \sin\theta_{imp} \approx 5.1$ cm/s for the formation of EERs.

An extremely elongated shape of the third largest remnant is formed through the impact with $\theta_{imp} = 15°$ and $v_{imp} = 24\,\mathrm{cm/s}$ as shown in Fig. 5.4. The impact has the larger impact velocity compared to that in Fig. 5.2, which leads to the formation of a large ejecta curtain (Fig. 5.4b). Then reaccumulation of fragments produces a filamentary structure (Fig. 5.4c), and further gravitational fragmentation of the

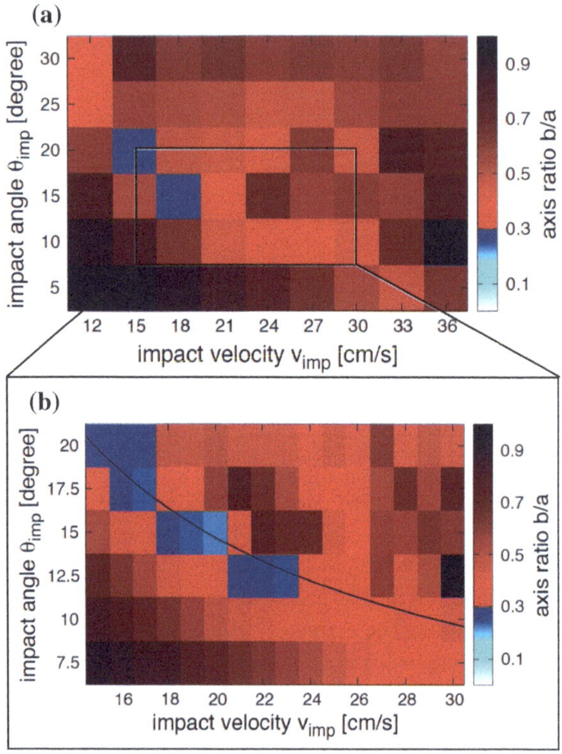

Fig. 5.3 Axis ratio b/a of the largest remnants produced through various equal-mass impacts ($q = 1$) with $\phi_d = 40°$, various v_{imp} (horizontal axis), and various θ_{imp} (vertical axis). The blue regions show impact parameters that form the largest remnants with $b/a < 0.3$, i.e., EERs. The black curve in the panel **b** shows $v_{imp} \sin\theta_{imp} = 5.1\,\mathrm{cm/s}$. Reprinted from Sugiura et al. (2019), Copyright (2019), with permission from Elsevier

filament forms the EER (Fig. 5.4d, e). The EER has the mass $0.24M_{target}$ and the axis ratio $b/a = 0.27$. However, the other EERs are obtained from the largest remnants.

5.3.2 Dependence on the Friction Angle

Figure 5.5 shows b/a of the largest remnants produced through various impacts with $q = 1$ and $\phi_d = 50°$. Note that b/a does not seem to be smooth function of v_{imp} and θ_{imp} because of a more rough parameter survey than that of Fig. 5.3b. Impacts with a wider parameter space produce EERs compared to impacts with $\phi_d = 40°$. We find that the parameters producing EERs concentrate around $v_{imp} \sin\theta_{imp} \approx 7.4\,\mathrm{cm/s}$ with $\theta_{imp} \leq 30°$ and $v_{imp} \leq 35\,\mathrm{m/s}$. Note that four impacts with $\theta_{imp} \approx 25°$, $v_{imp} \approx 20 - 30\,\mathrm{cm/s}$, and $\phi_d = 50°$ produce EERs from smaller remnants.

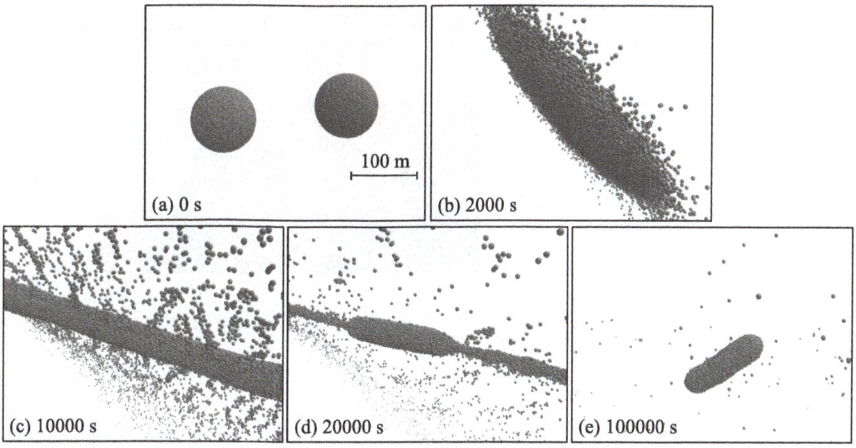

Fig. 5.4 Snapshots of the third largest remnant formed through the impact with $q = 1$, $\theta_{\rm imp} = 15°$, $v_{\rm imp} = 24\,{\rm cm/s}$, and $\phi_d = 40°$. Reprinted from Sugiura et al. (2019), Copyright (2019), with permission from Elsevier

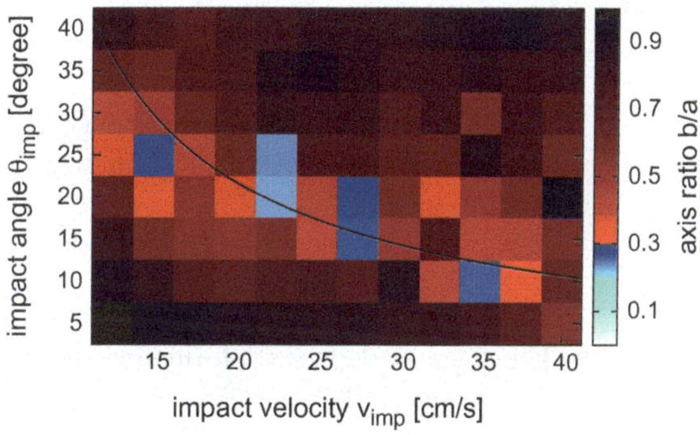

Fig. 5.5 b/a of the largest remnants produced through various equal-mass impacts ($q = 1$) with $\phi_d = 50°$. The black curve shows $v_{\rm imp} \sin\theta_{\rm imp} = 7.4\,{\rm cm/s}$. Reprinted from Sugiura et al. (2019), Copyright (2019), with permission from Elsevier

We also conduct impact simulations with $q = 1$ and $\phi_d = 30°$ (Fig. 5.6). Although we conduct impact simulations for the similar range of impact velocities and angles as in the parameter surveys with $\phi_d = 40°$ and $50°$ (Figs. 5.3 and 5.5), no impacts with $\phi_d = 30°$ produce EERs. Thus we require the friction angle $\phi_d \geq 40°$ for the formation of EERs.

We roughly estimate the critical friction angle required for the formation of EERs theoretically. As shown in Figs. 5.2 or 5.4, EERs are formed through collisional elongation of bodies and reaccumulation of fragments from directions of long axes. In reaccumulation phase, if the friction angle is small, the friction cannot stop fragments with small displacement, and resultant shapes become more rounded shapes. Total kinetic energy of fragments that compose an EER during reaccumulation phase is estimated as

$$E_{\rm kin} = \frac{1}{2} M_r v_{{\rm esc},r}^2, \tag{5.1}$$

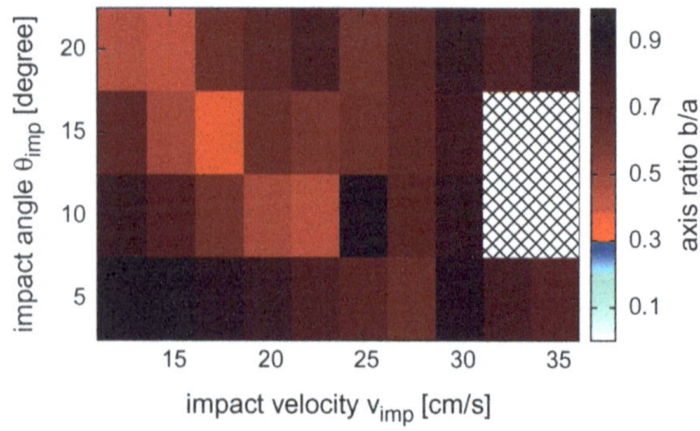

Fig. 5.6 b/a of the largest remnants produced through various equal-mass impacts ($q = 1$) with $\phi_d = 30°$. The cross-hatched region shows the impacts with $M_{\rm lr} < 0.15 M_{\rm target}$, and thus the shapes resulting from these impacts are not measured

where M_r is the mass of this EER and $v_{{\rm esc},r}$ is the escape velocity from this EER. According to Sugiura et al. (2018), total dissipated energy due to the friction with displacement L is estimated as

$$E_{\rm dis} = \pi R_r^2 L \mu_d P_c,$$
$$P_c = \frac{2}{3}\pi G \rho_0^2 R_r^2, \tag{5.2}$$

where G, R_r, and P_c are the gravitational constant, mean radius of the EER, and central pressure of the EER, respectively. Equating $E_{\rm kin}$ and $E_{\rm dis}$, we achieve

$$\frac{L}{R_r} \approx \frac{1}{\mu_d}. \tag{5.3}$$

If $L/R_r \gg 1$, i.e., $\mu_d \ll 1$, large deformation due to reaccumulation of fragments prevents the formation of EERs. Thus we roughly require μ_d larger than unity, or ϕ_d larger than 45°, for the formation of EERs.

Impacts with larger friction angles generally form more irregular shapes. We found one impact simulation that forms an EER with $b/a = 0.16$. Figure 5.7 shows the result of the impact simulation with $\theta_{\rm imp} = 15°$, $v_{\rm imp} = 22.5\,{\rm cm/s}$, and the friction angle $\phi_d = 60°$. The snapshots in Fig. 5.7 are similar to those in Fig. 5.2, and the impact with the slightly larger impact velocity and the slightly larger friction angle forms the more elongated shape with $b/a = 0.16$, which explains the axis ratio of 1I/'Oumuamua estimated in Bannister et al. (2017) or Fraser et al. (2018). Although the friction angle of $\phi_d = 60°$ is too large for usual terrestrial granular material (Heiken et al. 1991), granular material on small bodies may achieve an angle of repose of 60° owing to existence of cohesion.

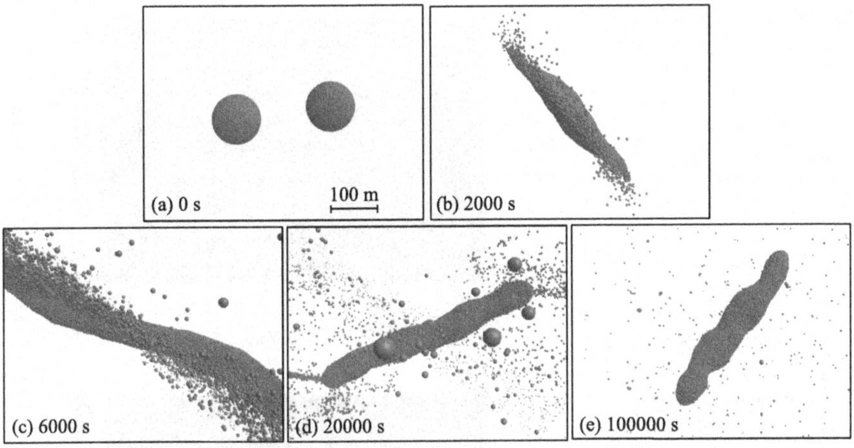

Fig. 5.7 Snapshots of the impact simulation with $q = 1$, $\theta_{\mathrm{imp}} = 15°$, $v_{\mathrm{imp}} = 22.5\,\mathrm{cm/s}$, and $\phi_d = 60°$

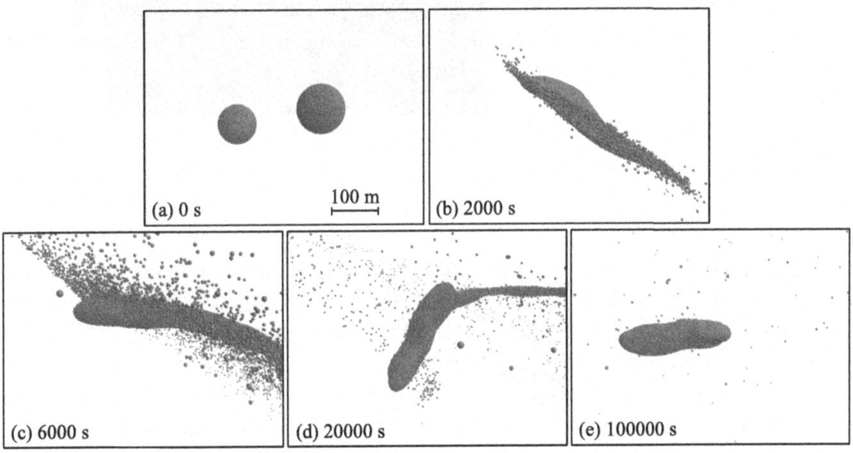

Fig. 5.8 Snapshots of the largest remnant formed through the impact with $q = 0.5$, $\theta_{\mathrm{imp}} = 20°$, $v_{\mathrm{imp}} = 25\,\mathrm{cm/s}$, and $\phi_d = 50°$. Reprinted from Sugiura et al. (2019), Copyright (2019), with permission from Elsevier

5.3.3 *Dependence on the Mass Ratio*

Figure 5.8 shows snapshots of the unequal-mas impact with $q = 0.5$, $\phi_d = 50°$, $\theta_{\mathrm{imp}} = 20°$, and $v_{\mathrm{imp}} = 25\,\mathrm{cm/s}$. Asymmetric configuration of the impact leads to asymmetric destruction of the bodies (Fig. 5.8b). Elongation of the target body is induced by the collision (Fig. 5.8c), which produces an EER (Fig. 5.8d, e). The EER has the mass of $1.1M_{\mathrm{target}}$ and the axis ratio b/a of 0.27.

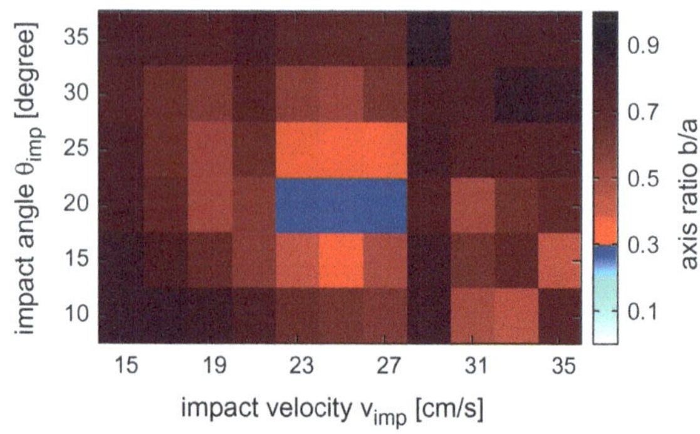

Fig. 5.9 b/a of the largest remnants produced through various unequal-mass impacts with $q = 0.5$ and $\phi_d = 50°$. Reprinted from Sugiura et al. (2019), Copyright (2019), with permission from Elsevier

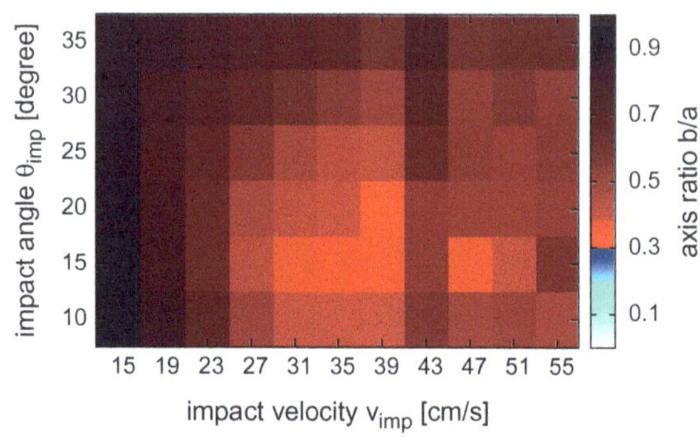

Fig. 5.10 b/a of the largest remnants produced through various unequal-mass impacts with $q = 0.25$ and $\phi_d = 50°$

Figure 5.9 shows b/a of the largest remnants produced through various impacts with $q = 0.5$ and $\phi_d = 50°$. Although it is more difficult to produce extreme shapes compared to the equal-mass impacts, we find that three impacts with $\theta_{\text{imp}} = 20°$ and $v_{\text{imp}} \approx 25\,\text{cm/s}$ produce EERs.

We also conduct impact simulations with $q = 0.25$ and $\phi_d = 50°$ (Fig. 5.10). We vary impact velocities in wider range, and the impacts investigated here include various types of collisions, i.e., merging collisions, hit-and-run collisions, and catastrophic destruction. However, no impacts with $q = 0.25$ produce EERs. Since sizes of deformed parts of target bodies are comparable to diameters of impactors, overall deformation of target bodies requires $q \sim 1$. Therefore we require $q \geq 0.5$ for the formation of EERs.

In summary, the formation of extremely elongated shapes with $b/a < 0.3$ roughly requires the mass ratio $q \geq 0.5$, the impact angle $\theta_{\text{imp}} \leq 30°$, the impact velocity $v_{\text{imp}} \leq 40\,\text{cm/s}$, and the friction angle $\phi_d \geq 40°$. Although we ignore cohesion forces in our simulations, it should be noted that the cohesion may increase required impact velocities for the formation of EERs. Shear strength due to the cohesion is at most 1 kPa (estimated from the cohesion of lunar soil: Heiken et al. 1991). In contrast, shear strength due to the friction is determined by the compression pressure, which is about 10 Pa in impacts between 100 m-sized bodies. Thus the cohesion may play an important role for minor-body impacts, and the cohesion may enlarge the range of the required impact velocities for the formation of EERs. However, the cohesion may prevent the elongation due to its large tensile strength. Thus how the cohesion affects shapes of collisional outcomes is investigated in our future works.

5.4 Discussion: Environments for 1I/'Oumuamua Formation

We need impact velocities of $15\,\mathrm{cm/s} \le v_{\mathrm{imp}} \le 40\,\mathrm{cm/s}$ for the collisional formation of EERs (see Figs. 5.3, 5.5, and 5.9), and such impact velocities are much smaller than typical relative velocities between objects orbiting around stars (e.g., the mean impact velocity at the main belt is about 5 km/s). To realize such small impact velocities, dynamically cold environments in protoplanetary disks are necessary. However, even in protoplanetary disks, relative velocities may increase larger than 40 cm/s due to perturbation caused by turbulence or gravitational interaction with large bodies. Here, we consider dynamical excitation of small bodies, or planetesimals, due to turbulence and large bodies, and we give the constraint on turbulent strengths and sizes of large bodies such that impact velocities become smaller than 40 cm/s.

5.4.1 *Turbulence*

We estimate relative velocity between planetesimals with the radius of $s = 100\,\mathrm{m}$ and the density of $\rho_s = 1\,\mathrm{g/cm^3}$ in a protoplanetary disk with a central star that has the same mass $M_\odot$ and luminosity as the Sun. The sound speed $C_{s,g}$, the density at the mid plane ρ_g, and the surface density Σ_g of gas at orbital radius r are given by those of the minimum mass solar nebula model (Hayashi 1981):

$$C_{s,g} = 1.0 \times 10^5 \left(\frac{r}{1\,\mathrm{au}}\right)^{-1/4} \mathrm{cm/s},$$

$$\rho_g = 1.4 \times 10^{-9} \left(\frac{r}{1\,\mathrm{au}}\right)^{-11/4} \mathrm{g/cm^3}, \tag{5.4}$$

$$\Sigma_g = 1.7 \times 10^3 \left(\frac{r}{1\,\mathrm{au}}\right)^{-3/2} \mathrm{g/cm^2}. \tag{5.5}$$

Planetesimals are dragged by gas. A time scale of velocity change due to gas drag is called the stopping time t_s and defined by the ratio of momentum to gas drag force. t_s for various gas drag laws is expressed as

$$t_s = \begin{cases} \frac{\rho_s s}{\rho_g \bar{v}_{\rm th}} & \text{for } s < \frac{9}{4}\lambda \quad \text{(Epstein law)} \\ \frac{4\rho_s s^2}{9\rho_g \lambda \bar{v}_{\rm th}} & \text{for } R_e < 1 \quad \text{(Stokes law 1)} \\ \frac{4^{3/5}\rho_s s^{8/5}}{9\rho_g \lambda^{3/5} \bar{v}_{\rm th}^{3/5} \Delta v^{3/5}} & \text{for } 1 < R_e < 800 \quad \text{(Stokes law 2)} \\ \frac{6\rho_s s}{\rho_g \Delta v} & \text{for } R_e > 800 \quad \text{(Newton law)} \end{cases}, \tag{5.6}$$

where $\bar{v}_{\rm th} = \sqrt{\pi/8}C_{s,g}$ is the mean thermal velocity of molecules, $\lambda = m_{\rm H_2}/\rho_g \sigma_{\rm H_2}$ is the mean free path of gas particles, $R_e = 4s\Delta v/\bar{v}_{\rm th}\lambda$ is the Reynolds number, Δv shows relative velocity between gas and planetesimals, $m_{\rm H_2} = 3.4 \times 10^{-24}$ g is the mass of a hydrogen molecule, and $\sigma_{\rm H_2} = 2.0 \times 10^{-15} {\rm cm}^2$ is the cross section of a hydrogen molecule (e.g., Whipple 1972; Weidenschilling 1977; Birnstiel et al. 2010). In protoplanetary disks, Δv is mainly determined by the difference between the Keplerian velocity $v_{\rm K} = \sqrt{GM_\odot/r}$ and the rotation velocity of gas with pressure-gradient support. From the relation of isothermal gas pressure $P = \rho_g C_{s,g}^2$ and Eq. (5.5),

$$\Delta v = \frac{1}{2}\frac{r}{\rho_g v_{\rm K}}\frac{dP}{dr} = 5.4 \times 10^3 \ {\rm cm/s}, \tag{5.7}$$

and λ and R_e are expressed as

$$\lambda = 1.2\left(\frac{r}{1\ {\rm au}}\right)^{11/4} {\rm cm},$$
$$R_e = 2.9 \times 10^3 \left(\frac{r}{1\ {\rm au}}\right)^{-5/2}. \tag{5.8}$$

From Eqs. (5.5), (5.7), and (5.8), t_s as a function of r is expressed as

$$t_s = \begin{cases} 2.2 \times 10^1 \left(\frac{r}{1\,{\rm AU}}\right)^{3/2} \Omega_{\rm K}^{-1} & \text{for } r > 19\,{\rm AU} \\ 3.6 \times 10^3 \left(\frac{r}{1\,{\rm AU}}\right)^{-1/4} \Omega_{\rm K}^{-1} & \text{for } 1.6\,{\rm AU} < r < 19\,{\rm AU} \\ 1.8 \times 10^3 \left(\frac{r}{1\,{\rm AU}}\right)^{5/4} \Omega_{\rm K}^{-1} & \text{for } r < 1.6\,{\rm AU} \end{cases}, \tag{5.9}$$

where $\Omega_{\rm K}$ is the Keplerian angular frequency. The gas drag is determined by the Epstein law for $r > 19$ AU, the Stokes law for 1.6 AU $< r <$ 19 AU, and the Newton law (high Reynolds number) for $r <$ 1.6 AU. As we notice from Eq. (5.9), $t_s\Omega_{\rm K} \gg 1$ for all regimes of the protoplanetary disk, i.e., the stopping timescale is much longer than the Keplerian rotation period. We estimate the relative velocity $v_{\rm rel}$ determined by turbulent stirring with such minor gas drag forces as follows (Cuzzi et al. 2001; Ormel and Cuzzi 2007):

$$v_{\rm rel} \approx C_{s,g}\sqrt{3\alpha(\Omega_{\rm K}t_s)^{-1}}, \tag{5.10}$$

where α is a parameter that represents the turbulent strength in Shakura-Sunyaev prescription. Figure 5.11 shows $v_{\rm rel}$ as a function of r and α, and we find that

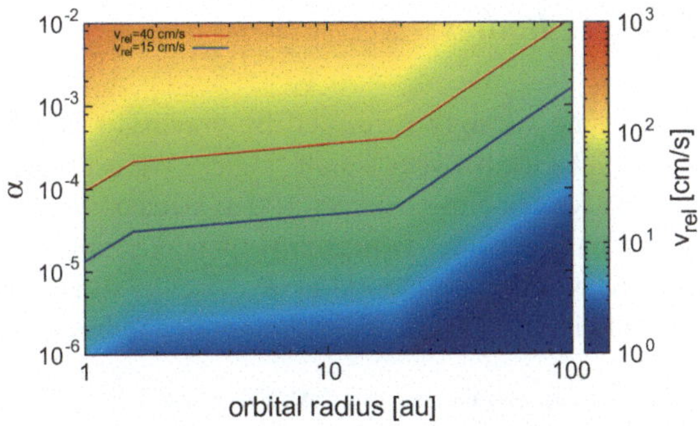

Fig. 5.11 Relative velocity between 100 m-sized planetesimals stirred up by turbulence in the minimum mass solar nebula disk as a function of the orbital radius (horizontal axis) and the turbulent strength α (vertical axis). The red and blue solid lines show the contours of $v_{\rm rel} = 40\,{\rm cm/s}$ and $v_{\rm rel} = 15\,{\rm cm/s}$, respectively. Reprinted from Sugiura et al. (2019), Copyright (2019), with permission from Elsevier

low-velocity impacts with $v_{\rm imp} \leq 40\,{\rm cm/s}$ are realized with $\alpha < 5 \times 10^{-4}$ for $1\,{\rm AU} < r < 20\,{\rm AU}$ and $\alpha < 5 \times 10^{-4}(r/20\,{\rm AU})^2$ for $r > 20\,{\rm AU}$.

In protoplanetary disks, turbulence is mainly driven by Magnetorotational instability (MRI). At mid planes of protoplanetary disks, MRI is active for $r < 0.1$ AU (vicinity of central stars) because of high ionized fraction and for $r > 10$ AU (sufficiently distant places from central stars) because of low surface density of gas. In such MRI active regions, strong turbulence with $\alpha \sim 10^{-2}$ is induced (e.g., Flock et al. 2017). In contrast, MRI is not active for $0.1\,{\rm AU} < r < 10\,{\rm AU}$ (dead zones: e.g., Kretke et al. 2009). The turbulent strength in dead zones is estimated to be $10^{-6} < \alpha < 10^{-3}$ (e.g., Flock et al. 2017; Ilgner and Nelson 2008; Mori et al. 2017). Thus, 100 m-sized bodies probably collide each other with low velocities of $v_{\rm imp} \leq 40\,{\rm cm/s}$ in dead zones.

α for several protoplanetary disks seems to be smaller than the value expected from MRI turbulence even for outer parts of the disks. For example, the protoplanetary disk HL Tau shows clear ring-like gaps with the radius $\sim$ 100 au. For such appearance of the gaps, HL Tau should be thin disk, which requires $\alpha \sim 10^{-4}$ (Pinte et al. 2016). Note that multiple ring-like structure observed in HL Tau can be produced through secular gravitational instability and small values of $\alpha < 10^{-4}$ is necessary for the occurrence of the secular gravitational instability (Takahashi and Inutsuka 2014, 2016; Tominaga et al. 2018). Thus impact velocities of $v_{\rm imp} \leq 40\,{\rm cm/s}$ are realized in weak turbulent protoplanetary disks such as HL Tau.

5.4.2 Size of Larger Bodies

Orbits of small bodies are strongly perturbed by gravity of larger bodies, which increases the relative velocity between small bodies. We estimate sizes of larger bodies that result in relative velocities between 100 m-sized bodies less than 40 cm/s.

For simplicity, we only consider two groups of bodies: small bodies with the radius 100 m and large bodies with the radius R. Viscous stirring due to the large bodies increases the relative velocity between the smaller bodies u with a rate of (Ida and Makino 1993)

$$\left.\frac{1}{u}\frac{du}{dt}\right|_{\mathrm{vs}} \sim +\Omega_{\mathrm{K}}\frac{\Sigma}{\rho_s R}\left(\frac{v_{\mathrm{esc}}}{u}\right)^4, \tag{5.11}$$

where Σ represent the surface mass density of the large bodies and v_{esc} is the escape velocity from a large body. Gas drag forces decrease the relative velocity u with a rate of

$$\left.\frac{1}{u}\frac{du}{dt}\right|_{\mathrm{gd}} \sim -\frac{1}{t_s}, \tag{5.12}$$

which is given by the inverse of the stopping time. The balance between the increase of u due to gravitational scattering of the large bodies Eq. (5.11) and the decrease of u due to gas drag forces Eq. (5.12) determines the equilibrium relative velocity of

$$u = \left(\frac{t_s\Omega_{\mathrm{K}}\Sigma}{\rho_s R}\right)^{1/4} v_{\mathrm{esc}}. \tag{5.13}$$

Figure 5.12 shows the equilibrium relative velocity given by Eq. (5.13). Here, we put 10^{-2} times the solid surface density for Σ. From Fig. 5.12, we find that low-velocity impacts with $u < 40\,\mathrm{cm/s}$ require $R < 2\,\mathrm{km}$ for $r \approx 1$ AU and $R < 7\,\mathrm{km}$ for $r > 20$ AU. Therefore, the formation of the extremely elongated shape of 1I/'Oumuamua through a collision requires an extremely young protoplanetary disk without bodies larger than 7 km.

We discuss the possibility that the shape of 1I/'Oumuamua is formed in a protoplanetary disk outside of the solar system. Thus, 1I/'Oumuamua should be ejected from the protoplanetary disk to come to the solar system after the formation of the shape. However, the ejection requires large scattering sources such as gas giants or central binary stars, which significantly increase the relative velocity between planetesimals. Therefore, the extremely elongated shape of 1I/'Oumuamua was probably formed in a place without larger bodies such as outer part of the protoplanetary disk, and then 1I/'Oumuamua drifted inward to the vicinity of gas giants or central binary stars to be ejected (e.g., Raymond et al. 2011, 2018; Jackson et al. 2018). Instead, 1I/'Oumuamua may be directly ejected from outer part of the disk due to an early stellar encounter (e.g., Kobayashi et al. 2005). Volatile substances of small bodies are removed when they closely encounter with stars, and thus ejection by central binary stars may be more favorable than the ejection by gas giants or stellar encounters

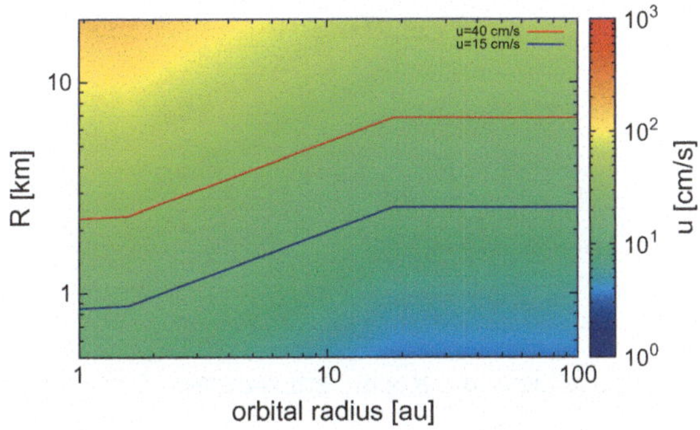

Fig. 5.12 Relative velocity between 100 m-sized planetesimals scattered by large bodies in the minimum mass solar nebula disk as a function of the orbital radius (horizontal axis) and the radius of large bodies R (vertical axis). The red and blue solid curves show the contours of $u = 40\,\mathrm{cm/s}$ and $u = 15\,\mathrm{cm/s}$, respectively. Reprinted from Sugiura et al. (2019), Copyright (2019), with permission from Elsevier

because the close encounter with central binary stars may explain lack of cometary activity of 1I/‘Oumuamua (Jackson et al. 2018).

5.4.3 Survivability of Extremely Elongated Shapes Through Ejection Processes

When objects closely encounter with scattering sources, strong tidal forces act on the objects. Then the objects may be disrupted due to the tidal forces. Especially, extremely elongated objects are more easily disrupted. Here, we discuss the survivability of the extremely elongated shape of 1I/‘Oumuamua through ejection processes.

We estimate the Roche radius of a scattering source with the density ρ_m, the radius R_M, and the mass $M = (4/3)\pi R_M^3 \rho_m$ when a prolate object with the major axis length a, the intermediate axis length b, and the density ρ_m closely encounters with the scattering source. We consider the situation that the major axis of the prolate body is pointing toward the scattering body because the prolate body is the most efficiently deformed in this situation. Tidal force per unit mass at the edges of the major axis of the prolate body is

$$F_t = \frac{GMa}{d^3}, \tag{5.14}$$

where d is the distance between the prolate body and the scattering body when the close encounter. According to Chandrasekhar (1969), self-gravitational force per unit mass at the edges of the major axis is calculated as

$$F_s = \pi G \rho_m A_1 a, \tag{5.15}$$

where

$$\begin{aligned} A_1 &= \frac{1-e^2}{e^3} \ln\left(\frac{1+e}{1-e}\right) - 2\frac{1-e^2}{e^2}, \\ e &= \sqrt{1-\frac{b^2}{a^2}}. \end{aligned} \tag{5.16}$$

Equating F_t and F_s, we achieve the Roche radius d_{R} as

$$d_{\mathrm{R}} = R_M \sqrt[3]{\frac{4}{3} A_1^{-1}}. \tag{5.17}$$

Even for an extreme shape with $b/a = 0.1$, $d_{\mathrm{R}} = 3.2 R_M$; the Roche radius is comparable to the radius of the scattering body.

Binary stars eject small bodies from inside of the critical orbital semi-major axis a_c (Holman and Wiegert 1999). An equal-mass binary with zero orbital eccentricity and orbital separation of a_b has $a_c = 2.4 a_b$. Since a_b is au-scale, a_c is much larger than radii of stars; small bodies are ejected before they experience tidal disruption.

In protoplanetary disks, planetesimals drift inward due to angular momentum loss through gas drag. Orbital eccentricities of planetesimals that drift to the vicinity of gas giants increase through moderate close encounters with gas giants. Distances of the close encounters required for the increase of the eccentricities are about the Hill radius (Ohtsuki et al. 2002), which is much larger than the Roche radius. The planetesimals with the eccentricities larger than 0.4 are efficiently ejected by the gas giants (Bandermann and Wolstencroft 1971; Higuchi et al. 2006), and typical distances of close encounters that eject planetesimals are also about the Hill radius (Fernandez 1978; Ida 1990). In the case of Jupiter, Hill radius is about 760 times larger than the Jupiter radius, and thus planetesimals rarely experience tidal disruption when they are ejected by gas giants.

References

L.W. Bandermann, R.D. Wolstencroft, Energy changes in close planetary encounters. Mon. Not. R. Astron. Soc. **152**, 377 (1971). https://doi.org/10.1093/mnras/152.4.377

M.T. Bannister, M.E. Schwamb, W.C. Fraser, M. Marsset, A. Fitzsimmons, S.D. Benecchi, P. Lacerda, R.E. Pike, J.J. Kavelaars, A.B. Smith, S.O. Stewart, S.-Y. Wang, M.J. Lehner, Col-OSSOS: colors of the interstellar planetesimal 1I/’Oumuamua. Astrophys. J. **851**, L38 (2017). https://doi.org/10.3847/2041-8213/aaa07c

W. Benz, E. Asphaug, Catastrophic disruptions revisited. Icarus **142**, 5–20 (1999). https://doi.org/10.1006/icar.1999.6204

T. Birnstiel, C.P. Dullemond, F. Brauer, Gas- and dust evolution in protoplanetary disks. Astron. Astrophys. **513**, A79 (2010). https://doi.org/10.1051/0004-6361/200913731

B.T. Bolin, H.A. Weaver, Y.R. Fernandez, C.M. Lisse, D. Huppenkothen, R.L. Jones, M. Jurić, J. Moeyens, C.A. Schambeau, C.T. Slater, Ž. Ivezić, A.J. Connolly, APO time-resolved color photometry of highly elongated interstellar object 1I/'Oumuamua. Astrophys. J. **852**, L2 (2018). https://doi.org/10.3847/2041-8213/aaa0c9

W.F. Bottke, D.D. Durda, D. Nesvorný, R. Jedicke, A. Morbidelli, D. Vokrouhlický, H. Levison, The fossilized size distribution of the main asteroid belt. Icarus **175**, 111–140 (2005). https://doi.org/10.1016/j.icarus.2004.10.026

S. Chandrasekhar, *Ellipsoidal Figures of Equilibrium* (1969)

J.N. Cuzzi, R.C. Hogan, J.M. Paque, A.R. Dobrovolskis, Size-selective concentration of chondrules and other small particles in protoplanetary nebula turbulence. Astrophys. J. **546**, 496–508 (2001). https://doi.org/10.1086/318233

C. de la Fuente Marcos, R. de la Fuente Marcos, Pole, pericenter, and nodes of the interstellar minor body A/2017 U1. Res. Notes AAS **1**, 5 (2017). https://doi.org/10.3847/2515-5172/aa96b4

G. Domokos, A.Á. Sipos, G.M. Szabó, P.L. Várkonyi, Formation of sharp edges and planar areas of asteroids by polyhedral abrasion. Astrophys. J. **699**(1), L13 (2009)

G. Domokos, A.Á. Sipos, G.M. Szabó, P.L. Várkonyi, Explaining the elongated shape of Oumuamua by the eikonal abrasion model. Res. Notes AAS **1**(1), 50 (2017)

M. Drahus, P. Guzik, W. Waniak, B. Handzlik, S. Kurowski, S. Xu, Tumbling motion of 1I/'Oumuamua and its implications for the body's distant past. Nature **2**, 407–412 (2018). https://doi.org/10.1038/s41550-018-0440-1

J.A. Fernandez, Mass removed by the outer planets in the early solar system. Icarus **34**, 173–181 (1978). https://doi.org/10.1016/0019-1035(78)90134-3

M. Flock, S. Fromang, N.J. Turner, M. Benisty, 3D radiation nonideal magnetohydrodynamical simulations of the inner rim in protoplanetary disks. Astrophys. J. **835**, 230 (2017). https://doi.org/10.3847/1538-4357/835/2/230

W.C. Fraser, P. Pravec, A. Fitzsimmons, P. Lacerda, M.T. Bannister, C. Snodgrass, I. Smoli'c, The tumbling rotational state of 1I/'Oumuamua. Nature **2** (2018). https://doi.org/10.1038/s41550-018-0398-z

C. Hayashi, Structure of the solar nebula, growth and decay of magnetic fields and effects of magnetic and turbulent viscosities on the nebula. Prog. Theor. Phys. Suppl. **70**, 35–53 (1981). https://doi.org/10.1143/PTPS.70.35

G.H. Heiken, D.T. Vaniman, B.M. French, *Lunar Sourcebook - A User's Guide to the Moon* (1991)

A. Higuchi, E. Kokubo, T. Mukai, Scattering of planetesimals by a planet: formation of comet cloud candidates. Astron. J. **131**, 1119–1129 (2006). https://doi.org/10.1086/498892

M.J. Holman, P.A. Wiegert, Long-term stability of planets in binary systems. Astron. J. **117**, 621–628 (1999). https://doi.org/10.1086/300695

S. Ida, Stirring and dynamical friction rates of planetesimals in the solar gravitational field. Icarus **88**, 129–145 (1990). https://doi.org/10.1016/0019-1035(90)90182-9

S. Ida, J. Makino, Scattering of planetesimals by a protoplanet - slowing down of runaway growth. Icarus **106**, 210 (1993). https://doi.org/10.1006/icar.1993.1167

M. Ilgner, R.P. Nelson, Turbulent transport and its effect on the dead zone in protoplanetary discs. Astron. Astrophys. **483**, 815–830 (2008). https://doi.org/10.1051/0004-6361:20079307

A.P. Jackson, D. Tamayo, N. Hammond, M. Ali-Dib, H. Rein, Ejection of rocky and icy material from binary star systems: Implications for the origin and composition of 1I/'Oumuamua. Mon. Not. R. Astron. Soc. (2018). https://doi.org/10.1093/mnrasl/sly033

D. Jewitt, J. Luu, J. Rajagopal, R. Kotulla, S. Ridgway, W. Liu, T. Augusteijn, Interstellar interloper 1I/2017 U1: observations from the NOT and WIYN telescopes. Astrophys. J. **850**(2), L36 (2017)

M. Jutzi, E. Asphaug, The shape and structure of cometary nuclei as a result of low-velocity accretion. Science **348**, 1355–1358 (2015). https://doi.org/10.1126/science.aaa4747

M.M. Knight, S. Protopapa, M.S.P. Kelley, T.L. Farnham, J.M. Bauer, D. Bodewits, L.M. Feaga, J.M. Sunshine, On the rotation period and shape of the hyperbolic asteroid 1I/'Oumuamua (2017 U1) from its lightcurve. Astrophys. J. **851**(2), L31 (2017)

H. Kobayashi, S. Ida, H. Tanaka, The evidence of an early stellar encounter in Edgeworth Kuiper belt. Icarus **177**, 246–255 (2005). https://doi.org/10.1016/j.icarus.2005.02.017
K.A. Kretke, D.N.C. Lin, P. Garaud, N.J. Turner, Assembling the building blocks of giant planets around intermediate-mass stars. Astrophys. J. **690**, 407–415 (2009). https://doi.org/10.1088/0004-637X/690/1/407
Z.M. Leinhardt, D.C. Richardson, T. Quinn, Direct N-body simulations of rubble pile collisions. Icarus **146**, 133–151 (2000). https://doi.org/10.1006/icar.2000.6370
Z.M. Leinhardt, R.A. Marcus, S.T. Stewart, The formation of the collisional family around the dwarf planet haumea. Astrophys. J. **714**, 1789–1799 (2010). https://doi.org/10.1088/0004-637X/714/2/1789
K.J. Meech, R. Weryk, M. Micheli, J.T. Kleyna, O.R. Hainaut, R. Jedicke, R.J. Wainscoat, K.C. Chambers, J.V. Keane, A. Petric, L. Denneau, E. Magnier, T. Berger, M.E. Huber, H. Flewelling, C. Waters, E. Schunova-Lilly, S. Chastel, A brief visit from a red and extremely elongated interstellar asteroid. Nature **552**, 378–381 (2017). https://doi.org/10.1038/nature25020
M. Micheli, D. Farnocchia, K.J. Meech, M.W. Buie, O.R. Hainaut, D. Prialnik, N. Schörghofer, H.A. Weaver, P.W. Chodas, J.T. Kleyna, R. Weryk, R.J. Wainscoat, H. Ebeling, J.V. Keane, K.C. Chambers, D. Koschny, A.E. Petropoulos, Non-gravitational acceleration in the trajectory of 1I/2017 U1 (‘Oumuamua). Nature **559**, 223–226 (2018). https://doi.org/10.1038/s41586-018-0254-4
T. Michikami, A. Hagermann, T. Kadokawa, A. Yoshida, A. Shimada, S. Hasegawa, A. Tsuchiyama, Fragment shapes in impact experiments ranging from cratering to catastrophic disruption. *Icarus* **264**(Supplement C), 316–330 (2016). https://doi.org/10.1016/j.icarus.2015.09.038
S. Mori, T. Muranushi, S. Okuzumi, S.-I. Inutsuka, Electron heating and saturation of self-regulating magnetorotational instability in protoplanetary disks. Astrophys. J. **849**, 86 (2017). https://doi.org/10.3847/1538-4357/aa8e42
K. Ohtsuki, G.R. Stewart, S. Ida, Evolution of Planetesimal Velocities Based on Three-Body Orbital Integrations and Growth of Protoplanets. Icarus **155**, 436–453 (2002). https://doi.org/10.1006/icar.2001.6741
C.W. Ormel, J.N. Cuzzi, Closed-form expressions for particle relative velocities induced by turbulence. Astron. Astrophys. **466**, 413–420 (2007). https://doi.org/10.1051/0004-6361:20066899
C. Pinte, W.R.F. Dent, F. Ménard, A. Hales, T. Hill, P. Cortes, I. de Gregorio-Monsalvo, Dust and gas in the disk of HL Tauri: surface density, dust settling, and dust-to-gas ratio. Astrophys. J. **816**, 25 (2016). https://doi.org/10.3847/0004-637X/816/1/25
S.N. Raymond, P.J. Armitage, A. Moro-Martín, M. Booth, M.C. Wyatt, J.C. Armstrong, A.M. Mandell, F. Selsis, A.A. West, Debris disks as signposts of terrestrial planet formation. Astron. Astrophys. **530**, A62 (2011). https://doi.org/10.1051/0004-6361/201116456
S.N. Raymond, P.J. Armitage, D. Veras, E.V. Quintana, T. Barclay, Implications of the interstellar object 1I/‘Oumuamua for planetary dynamics and planetesimal formation. Mon. Not. R. Astron. Soc. **476**, 3031–3038 (2018). https://doi.org/10.1093/mnras/sty468
D.C. Richardson, Z.M. Leinhardt, H.J. Melosh, W.F. Bottke, Jr., E. Asphaug, Gravitational aggregates: evidence and evolution, 501–515 (2002)
D.C. Richardson, P. Elankumaran, R.E. Sanderson, Numerical experiments with rubble piles: equilibrium shapes and spins. Icarus **173**, 349–361 (2005). https://doi.org/10.1016/j.icarus.2004.09.007
K. Sugiura, H. Kobayashi, S. Inutsuka, Collisional elongation: possible origin of extremely elongated shape of 1I/‘Oumuamua. Icarus **328**, 14–22 (2019). https://doi.org/10.1016/j.icarus.2019.03.014
K. Sugiura, H. Kobayashi, S. Inutsuka, Toward understanding the origin of asteroid geometries, variety in shapes produced by equal-mass impacts. Astron. Astrophys. **620**, A167 (2018). https://doi.org/10.1051/0004-6361/201833227
S.Z. Takahashi, S.-I. Inutsuka, Two-component secular gravitational instability in a protoplanetary disk: a possible mechanism for creating ring-like structures. Astrophys. J. **794**, 55 (2014). https://doi.org/10.1088/0004-637X/794/1/55

S.Z. Takahashi, S.-I. Inutsuka, An origin of multiple ring structure and hidden planets in HL Tau: a unified picture by secular gravitational instability. Astron. J. **152**, 184 (2016). https://doi.org/10.3847/0004-6256/152/6/184

R.T. Tominaga, S.-I. Inutsuka, S.Z. Takahashi, Non-linear development of secular gravitational instability in protoplanetary disks. Publ. Astron. Soc. Japan **70**, 3 (2018). https://doi.org/10.1093/pasj/psx143

K.J. Walsh, D.C. Richardson, Binary near-earth asteroid formation: rubble pile model of tidal disruptions. Icarus **180**(1), 201–216 (2006). https://doi.org/10.1016/j.icarus.2005.08.015

S.J. Weidenschilling, Aerodynamics of solid bodies in the solar nebula. Mon. Not. R. Astron. Soc. **180**, 57–70 (1977). https://doi.org/10.1093/mnras/180.1.57

F.L. Whipple, On certain aerodynamic processes for asteroids and comets, in *From Plasma to Planet*, ed. by A. Elvius (1972), p. 211

Q.-Z. Ye, Q. Zhang, M.S.P. Kelley, P.G. Brown, 1I/2017 U1 (‘Oumuamua) is hot: imaging, spectroscopy, and search of meteor activity. Astrophys. J. **851**(1), L5 (2017)

Chapter 6
Summary and Future Prospects

Abstract We firstly developed a numerical simulation code for rocky body collisions. Then we investigated shapes of asteroids formed by various impacts through numerical simulations. We also analyzed shapes of actual asteroids and compare them with shapes of impact outcomes produced through our simulations. Finally we applied our simulations to the formation of the extremely elongated shape of 1I/'Oumuamua. In this chapter, we summarize our study. We also state future prospects of our study such as application to other phenomena.

Keywords Summary · Simulation method for rocky body impacts · Asteroid shapes formed through collisions · Application

6.1 Summary

Asteroids in the present solar system are fossils of planetesimals that existed in the primordial solar system. Thus information obtained from asteroids is very useful to reveal the history of the solar system, and such information is also used to reveal the collisional history of asteroids themselves, i.e., how impacts they experienced or when they were formed. Asteroids have variety of shapes and we can directly connect shapes of asteroids to the collisional history of asteroids since asteroidal shapes are mainly formed through collisions. To do so, we need to know shapes of asteroids produced by various impacts through numerical simulations.

6.1.1 Development of the Numerical Simulation Method

Our simulation method is based on elastic dynamics because rocky bodies have elastic properties when deformation is small. We developed our simulation code for rocky body impacts based on the SPH method for elastic dynamics because particle methods are suitable to compute impact phenomena. However, the standard SPH method has the serious problem in tension dominated regions, the tensile instability,

K. Sugiura, *Development of a Numerical Simulation Method for Rocky Body Impacts and Theoretical Analysis of Asteroidal Shapes*, Springer Theses,
https://doi.org/10.1007/978-981-15-3722-6_6

and causes numerical instability in tension dominated regions. We developed the new method to solve the tensile instability based on the Godunov SPH method (Inutsuka 2002). We evaluated the stability of the Godunov SPH method through the linear stability analysis and clarified the stability with several interpolation methods of density distribution as shown in Table 2.1 (Sugiura and Inutsuka 2016, 2017). We also found that we can conduct numerical simulations without the tensile instability if we select stable interpolation methods, which depend on the sign of pressure. Our test simulations show that the Godunov SPH method for elastic dynamics well reproduces the dynamics of elastic bodies (Figs. 2.4 and 2.5 or Sugiura and Inutsuka 2017).

However, asteroids are not elastic bodies but rocky bodies and we need appropriate methods to describe impacts of rocky bodies. The equation of state that is appropriate to describe impact phenomena is different from that of elastic dynamics. Many studies investigating impact phenomena with the SPH method utilize the Tillotson equation of state (e.g., Genda et al. 2012; Jutzi and Asphaug 2015), and thus we also utilize the Tillotson equation of state. Large deformation of rocky bodies causes the fracturing, which relieves tensile forces of intact rocks. We included the fracture model for brittle solids described in Benz and Asphaug (1995). Friction between completely fractured rocks, or granular materials, is very important to describe shapes of rubble pile bodies such as the asteroid Itokawa. Thus we included the friction model for granular materials described in Jutzi (2015). Our test simulations show that our simulation code well reproduces the dynamics of granular materials (Fig. 2.6). To reproduce the collisional formation of shapes of rubble pile bodies, we also need to describe reaccumulation of fragments ejected by impacts, and thus we included the self gravity.

Catastrophic collisions produce fine remnants. To reproduce and investigate shapes of such fine remnants, we need to conduct high-resolution simulations. To do so, we need to parallelize our simulation code to utilize supercomputers. Thus we parallelized our simulation code utilizing FDPS developed by RIKEN. As we notice from Fig. 2.7, our simulation code successfully achieves the high parallelization efficiency even if we utilize 2,000 CPU cores.

6.1.2 Asteroidal Shapes Formed Through Collisions

DAMIT provides shape models of asteroids obtained through light curve observations of asteroids, and almost all of shapes of asteroids with diameters larger than 100 km are already measured (Fig. 4.3). Moreover, the collisional lifetimes of asteroids larger than 100 km are much longer than the age of the solar system (O'Brien and Greenberg 2005), and many of them are not probably affected by recent impacts (Fig. 4.4). Therefore, the information of the asteroids larger than 100 km is desirable to investigate the long collisional history of those asteroids, and in this study we mainly focus on the shapes of the asteroids larger than 100 km.

6.1.2.1 Impact Velocity in the History of the Solar System

In the early planet formation era, planetesimals are under orderly growth until the largest bodies grow up to be size of about 100 km. In this stage, impact velocities between planetesimals are comparable to or slightly larger than escape velocities from planetesimals (Kobayashi et al. 2016). Then, several planetesimals larger than 100 km grow up to be protoplanets rapidly under runaway growth, and evolutionary stage moves on to oligarchic growth stage. The impact velocities between planetesimals are controlled by the mass of protoplanets, and the impact velocities are kept to be as low as several 100 m/s until the mass of protoplanets becomes that of Mars (see Eq. (4.6)). Since escape velocity from a planetesimal with diameter of 100 km is about 100 m/s, the impact velocities between 100 km-sized planetesimals are always comparable to the escape velocity until the mass of protoplanets reaches to that of Mars. Therefore, impacts in the primordial environment mainly result in non-destructive impacts.

In contrast, impact velocities between asteroids in the present solar system are much larger than escape velocities from asteroids because of the existence of gas giants. The average impact velocity in the main belt is about 5 km/s, and occurrence rate of impacts with impact velocities less than 400 m/s is only about 0.2% (Fig. 4.1). Therefore, impacts in the present environment mainly result in destructive impacts.

6.1.2.2 Shapes Formed Through Various Impacts

We investigated shapes of remnants produced through various impacts of asteroids using numerical simulations. We consider impacts between basaltic asteroids that initially have spherical shapes and no rotation. Diameter of a target asteroid, i.e., larger asteroid among two impacting asteroids, is fixed to be 100 km, and we vary impact velocities, impact angles, and mass ratios of two impacting asteroids.

Firstly, we investigated shapes of the largest remnants produced through non-destructive impacts that have the mass of the largest remnants larger than $0.4M_{\text{target}}$, where M_{target} is the mass of initial target asteroids. Note that fine remnants other than the largest remnants are also produced through non-destructive impacts, and we expect that shapes of such fine remnants are similar to those produced through destructive impacts because those shapes are formed through gravitational reaccumulation of fine fragments ejected by impacts, which is the same mechanism observed in destructive impacts. Our simulations of equal-mass impacts with impact velocities comparable to escape velocity show that these impacts produce various shapes including extremely flat shapes of the largest remnants (Fig. 3.15). However, we notice that impacts with smaller mass ratios of two impacting asteroids are more difficult to produce irregular shapes (Fig. 3.18), and impacts with the mass ratio of 1/64 only produce the spherical largest remnants with the ratio of the minor to major axis lengths larger than 0.7 (Fig. 3.19).

Secondly, we investigated shapes of fine remnants produced through destructive impacts by means of high-resolution simulations. We conducted four simulations of destructive impacts with various mass ratios and impact velocities. The highest impact velocity among four simulations is about 2 km/s, which is close to the average impact velocity in the present solar system. As a result, we found that the destructive impacts mainly produce spherical and bilobed shapes of fine remnants (Fig. 3.22). We also found that flat shapes are difficult to be formed through the destructive impacts (Fig. 3.24).

6.1.2.3 Shapes Formed in Primordial and Present Environment

Impact velocities between planetesimals in the planet formation era are comparable to escape velocities from planetesimals. Thus similar-mass impacts in the primordial environment forms various shapes including extremely flat shapes. Although similar mass impacts between 100 km-sized asteroids in the present solar system are very rare, similar-mass impacts between planetesimals in the primordial environment are more frequent compared to the present environment because Jupiter formation depleted planetesimals. Even in the primordial environment, impacts with large mass difference of two impacting planetesimals are more frequent than similar-mass impacts. However, low-velocity impacts with large mass difference do not result in the deformation of target planetesimals. Since 100 km-sized planetesimals are mainly the largest bodies among planetesimals in the primordial environment (Kobayashi and Tanaka 2018), shapes of 100 km-sized planetesimal are mainly determined by similar-mass impacts, which leads to the formation of various shapes including flat shapes.

Impact velocities between asteroids in the present solar system are much larger than escape velocities from asteroids, which mainly result in destructive impacts. Our simulations of destructive impacts show that these impacts only produce spherical and bilobed shapes. Even in the present environment, impacts with large mass difference of two impacting asteroids do not result in destructive impacts. However, our simulations with the mass ratio of 1/64 only produce the spherical largest remnants even with impact velocities realized in the present solar system. Therefore, we expect that impacts in the present solar system do not produce flat shapes of asteroids.

6.1.3 Comparison with Actual Asteroids

From the results of our simulations, we predict that destructive impacts that are realized in the present solar system are difficult to produce flat shapes. To verify this prediction, we investigated shapes of actual asteroids obtained from DAMIT. Asteroid families are mainly formed by recent destructive impacts, and thus we can assess the effect of destructive impacts on asteroidal shapes through the comparison between shapes of family and non-family asteroids. The information of asteroid families is obtained from AstDyS-2.

As a result of the investigation, we found that family asteroids with diameters larger than 100 km do not include flat asteroids, while non-family asteroids larger than 100 km include six flat asteroids with $c/a < 0.6$ and $b/a > 0.8$ (Fig. 4.6). This is consistent with the results of our simulations that destructive impacts do not produce flat shapes. Thus, the flat shapes of the asteroids with the diameters larger than 100 km were formed in the primordial environment and remain the same until today. This conclusion is consistent with the collisional lifetime estimated in O'Brien and Greenberg (2005).

In contrast, family asteroids with diameters larger than 10 km include flat asteroids. These flat asteroids in asteroid families may be formed through low-velocity collisions between similar-sized members in the same families. However, the fraction of flat asteroids with $c/a < 0.6$ and $b/a > 0.8$ shows the clear difference between family and non-family asteroids. The fraction of the flat shapes for the non-family asteroids is 8.5%, while that for the family asteroids is 3.8% (Fig. 4.8); the fraction of the flat shapes for the non-family asteroids is more than twice as large as that for the family asteroids. This difference also suggests that destructive impacts are difficult to produce flat asteroids.

6.1.4 Formation of Extremely Elongated Shape of 1I/'Oumuamua

As shown in Fig. 5.2, collisions can form extremely elongated shapes of asteroids. Recently, Meech et al. (2017) discovered the interstellar object 1I/'Oumuamua that has the extremely elongated shape. Owing to large amplitude of light curve of 1I/'Oumuamua, the axis ratio b/a of this object is estimated to be 0.3 or smaller, and the nominal value of b/a is considered to be 1/6. To investigate detailed impact conditions to form extremely elongated remnants (EERs) with $b/a < 0.3$ and size similar to that of 1I/'Oumuamua, we conducted impact simulations with 50 m-sized rubble pile targets. We varied impact velocities, impact angles, mass ratios of two impacting asteroids, and friction angles, and we searched for impact conditions to form EERs.

As a result, we found that the formation of EERs roughly requires the impact velocity smaller than 40 cm/s (Fig. 5.5), the impact angle smaller than 30° (Fig. 5.5), the mass ratio larger than 0.5 (Figs. 5.9 and 5.10), and the friction angle larger than 40° (Figs. 5.5 and 5.6).

The most difficult condition to be realized is the impact velocity. The realization of such low impact velocity requires dynamically extremely cold environment such as in a protoplanetary disk. We estimated required conditions of such protoplanetary disk in terms of turbulence and size of large planetesimals. As a result, we found that the impact velocity smaller than 40 cm/s is realized in the environments with Shakura-Sunyaev $\alpha < 10^{-4}$ such as in MRI dead zones (Fig. 5.11) and without planetesimals larger than 7 km (Fig. 5.12). Small size of planetesimals requires environments in an extremely young protoplanetary disk. Thus, 1I/'Oumuamua might have been formed in such an extremely young protoplanetary disk, then might have been ejected from the planetary system due to, for example, a stellar encounter, and then visited our solar system.

6.2 Future Prospects

6.2.1 Application to Other Phenomena

6.2.1.1 Cratering

In this study, we focus on collisions that deform overall structures of asteroids. However, as we notice from surfaces of terrestrial objects such as the moon, collisions also produce craters. Since collisions due to smaller objects are more frequent than those between similar-mass bodies, craters give us the opportunity to statistically discuss the collisional history in the solar system.

There are various types of craters. Small craters are generally simple craters that have bowl-like structures. Large craters have, in contrast, more complex structures such as central peaks or ring-like structures. Some celestial objects have craters that cover the large fraction of surfaces of objects. For example, Phobos, the satellite of Mars, has Stickney crater on its surface (e.g., Bruck Syal et al. 2016). The size of Stickney crater $\approx$9 km is comparable to the size of Phobos $\approx$22 km. Craters are not necessarily symmetric but some craters have asymmetric structures, which are considered to be formed by oblique impacts (Goeritz et al. 2009) or impacts to surface slopes (Krohn et al. 2014). These variety of crater types give us more detailed clues to reveal impact conditions that form craters or surface conditions of celestial bodies.

Although there are numerous works on the formations of craters through experimental (e.g., Housen and Holsapple 1999; Michikami et al. 2014) or numerical (e.g., Jutzi et al. 2010; Potter et al. 2018) methods, knowledge for the formations of complex craters or asymmetric craters is limited. The investigations of asymmetric craters through numerical simulations need three-dimensional simulations. However, three-dimensional simulations of crater formations are difficult because of high-computational costs. Since our simulation code can be applied to simulations of crater formations and are suitable for high-resolution simulations owing to the parallelization, the crater formation is a good target of our future works.

6.2.1.2 Tidal Disruption

Tidal disruption of asteroids due to close encounters with planets also changes asteroidal shapes. Some studies argue that the extremely elongated shape of 1I/'Oumuamua is formed through tidal disruption (e.g. Ćuk 2018). Although close encounters that cause tidal disruption are rare because of small size of Roche radius, the constraint on shapes of asteroids formed though tidal disruption leads to more detailed constraint on the origin of asteroids.

Walsh and Richardson (2006) well investigated properties of asteroids formed through tidal disruption in terms of the formation of binary asteroids through numerous simulations. However, shapes of remnants produced through tidal disruption are not well investigated so far. Since tidal disruption includes complex processes of approach to planets, disruption, and leaving from planets, numerical simulations are unique method to investigate tidal disruption. Our simulation code is applicable to tidal disruption problems easily.

6.2.1.3 Rotational Deformation

The spacecraft Hayabusa 2 recently clarified the diamond-like shape, or the top shape, of the asteroid Ryugu. There are other examples of asteroids that have top shapes. The target of OSIRIS-REx mission, the asteroid Bennu, also has the top shape. These top shapes are considered to be formed through rotational deformation due to the acceleration of rotation caused by the YORP effect (Walsh et al. 2008). The equatorial ridge of the asteroid Ryugu is also considered to be formed through the rotation of the asteroid.

Although Walsh et al. (2008) showed that top shapes are formed through rotations, we have not yet known the detailed dependence of internal structures or properties such as the friction angle of granular material on shapes of asteroids produced through rotations. Since shapes formed through rotations may depend on internal structures such as the existence of rigid cores or the values of friction angle or cohesion, detailed investigations of rotational deformation leads to not only constrain internal structures or properties of granular material but also constrain the formation processes of asteroids. Thus rotational deformation is one of the targets of our future works.

6.2.2 Addition of Other Models

6.2.2.1 Cohesion

The equation for shear strength of granular material is composed of two terms. One of them is the term that is proportional to confining pressure, and the other one is the constant term that corresponds to the shear strength at zero confining pressure. The latter term corresponds to the cohesion of granular material. Strength due to the cohesion is at most 1 kPa (Heiken et al. 1991), which is comparable to central pressure of 1 km-sized asteroids. For simulations of the collisional deformation of asteroids larger than 10 km, the cohesion is negligible because confining pressure due to the gravity is so large. However, the cohesion may play significant roles for simulations for asteroids smaller than 1 km or small structures such as mountains on asteroids. Especially, the cohesion probably affects the formation of extreme shapes of minor bodies such as the extremely elongated shape of 1I/'Oumuamua. We have not yet introduced the model of the cohesion to our simulation code, and thus we will include the cohesion referring to, for example, Alduán and Otaduy (2011), in our future works.

6.2.2.2 Porosity

Some rocky materials such as pumice have micro porosity, that is, they have void spaces inside of themselves. Materials that compose asteroids, e.g., chondrites, also have porosity. The mean value of porosity of chondrites is about 10% (Britt et al. 2002). This porosity may affect results of impacts because crushing of porous material dissipates impact energy.

Jutzi et al. (2008) introduced the effect of crushing of micro porosity to SPH simulations. Their porosity model is based on the P-α model (Herrmann 1969; Carroll and Holt 1972), where distension parameter α, or porosity, is determined by the pressure. Jutzi (2015) showed that the inclusion of the porosity model increases catastrophic disruption thresholds Q^*_{RD} by about factor two. Thus the porosity affects at least how impacts are destructive. They have not yet investigated the dependence of the porosity on shapes of asteroids formed through impacts, so that we will also introduce the porosity model to our simulation code.

6.2.3 Comparison with Particle Based Codes

We conducted simulations of catastrophic disruption only with the SPH method. However, simulation methods for fluid dynamics including the SPH method have a limitation of time steps, that is, the CFL condition. Time step determined by the CFL condition is generally much smaller than the timescale of gravitational reaccumulation. Thus, some researches utilize particle based codes to simulate gravitational reaccumulation phase (e.g., Michel and Richardson 2013; Schwartz et al. 2018).

Except for the self gravity, equations utilized in the SPH method and particle based codes are completely different. Moreover, the SPH method includes additional dissipation due to the artificial viscosity, while particle based codes do not include such additional dissipation. Although we expect that shapes of remnants produced in reaccumulation phase of catastrophic disruption are not affected by the difference of simulation methods because the shapes are mainly controlled by the self gravity, we should take note of the difference of simulation methods. We also should note that the difference of the definition of the friction angle. The friction angle in the SPH method means that of granular material, while the friction angle in particle based codes means that between modeled particles. This difference of the definition may cause the difference of shapes of produced remnants. In our future works, we will directly compare results of the same impact obtained from both simulation methods.

References

I. Alduán, M.A. Otaduy, Sph granular flow with friction and cohesion, in *Proceedings of the ACM SIGGRAPH/Eurographics Symposium on Computer Animation* (2011)

W. Benz, E. Asphaug, Simulations of brittle solids using smooth particle hydrodynamics. Comput. Phys. Commun. **87**, 253–265 (1995). https://doi.org/10.1016/0010-4655(94)00176-3

D.T. Britt, D. Yeomans, K. Housen, G. Consolmagno, *Asteroid Density, Porosity, and Structure* (2002), pp. 485–500

M. Bruck Syal, J. Rovny, J.M. Owen, P.L. Miller, Excavating Stickney crater at Phobos. Geophys. Res. Lett., **43**(20), 10,595–10,601 (2016). https://doi.org/10.1002/2016GL070749

M. Carroll, A.C. Holt, Suggested modification of the P-α model for porous materials. J. Appl. Phys. **43**, 759–761 (1972). https://doi.org/10.1063/1.1661203

M. Ćuk, 1I/'Oumuamua as a tidal disruption fragment from a binary star system. Astrophys. J. **852**, L15 (2018). https://doi.org/10.3847/2041-8213/aaa3db

H. Genda, E. Kokubo, S. Ida, Merging criteria for giant impacts of protoplanets. Astrophys. J. **744**, 137 (2012). https://doi.org/10.1088/0004-637X/744/2/137

M. Goeritz, T. Kenkmann, K. Wünnemann, S. van Gasselt, Asymmetric structure of lunar impact craters due to oblique impacts? in *Lunar and Planetary Science Conference*, vol. 40 (2009), p. 2096

G.H. Heiken, D.T. Vaniman, B.M. French, *Lunar sourcebook—A user's guide to the Moon* (1991)

W. Herrmann, Constitutive equation for the dynamic compaction of ductile porous materials. J. Appl. Phys. **40**, 2490–2499 (1969). https://doi.org/10.1063/1.1658021

K.R. Housen, K.A. Holsapple, Scale effects in strength-dominated collisions of rocky asteroids. Icarus **142**(1), 21–33 (1999). https://doi.org/10.1006/icar.1999.6206

S. Inutsuka, Reformulation of smoothed particle hydrodynamics with Riemann solver. J. Comput. Phys. **179**, 238–267 (2002). https://doi.org/10.1006/jcph.2002.7053

M. Jutzi, SPH calculations of asteroid disruptions: the role of pressure dependent failure models. Planet. Space Sci. **107**, 3–9 (2015). https://doi.org/10.1016/j.pss.2014.09.012

M. Jutzi, E. Asphaug, The shape and structure of cometary nuclei as a result of low-velocity accretion. Science **348**, 1355–1358 (2015). https://doi.org/10.1126/science.aaa4747

M. Jutzi, W. Benz, P. Michel, Numerical simulations of impacts involving porous bodies. I. Implementing sub-resolution porosity in a 3D SPH hydrocode. Icarus **198**, 242–255 (2008). https://doi.org/10.1016/j.icarus.2008.06.013

M. Jutzi, P. Michel, W. Benz, A large crater as a probe of the internal structure of the E-type asteroid Steins. Astron. Astrophys. **509**, L2 (2010). https://doi.org/10.1051/0004-6361/200913549

H. Kobayashi, H. Tanaka, From planetesimal to planet in turbulent disks. II. Formation of gas giant planets. Astrophys. J. **862**, 127 (2018). https://doi.org/10.3847/1538-4357/aacdf5

H. Kobayashi, H. Tanaka, S. Okuzumi, From planetesimals to planets in turbulent protoplanetary disks. I. Onset of runaway growth. Astrophys. J. **817**, 105 (2016). https://doi.org/10.3847/0004-637X/817/2/105

K. Krohn, R. Jaumann, D. Elbeshausen, T. Kneissl, N. Schmedemann, R. Wagner, J. Voigt, K. Otto, K.D. Matz, F. Preusker, T. Roatsch, K. Stephan, C.A. Raymond, C.T. Russell, Asymmetric craters on Vesta: impact on sloping surfaces. Planet. Space Sci. **103**, 36–56 (2014). https://doi.org/10.1016/j.pss.2014.04.011

K.J. Meech, R. Weryk, M. Micheli, J.T. Kleyna, O.R. Hainaut, R. Jedicke, R.J. Wainscoat, K.C. Chambers, J.V. Keane, A. Petric, L. Denneau, E. Magnier, T. Berger, M.E. Huber, H. Flewelling, C. Waters, E. Schunova-Lilly, S. Chastel, A brief visit from a red and extremely elongated interstellar asteroid. Nature **552**, 378–381 (2017). https://doi.org/10.1038/nature25020

P. Michel, D.C. Richardson, Collision and gravitational reaccumulation: possible formation mechanism of the asteroid Itokawa. Astron. Astrophys. **554**, L1 (2013). https://doi.org/10.1051/0004-6361/201321657

T. Michikami, A. Hagermann, H. Miyamoto, S. Miura, J. Haruyama, P. Lykawka, Impact cratering experiments in brittle targets with variable thickness: implications for deep pit craters on Mars. Planet. Space Sci. **96**, 71–80 (2014). https://doi.org/10.1016/j.pss.2014.03.010
D.P. O'Brien, R. Greenberg, The collisional and dynamical evolution of the main-belt and NEA size distributions. Icarus **178**, 179–212 (2005). https://doi.org/10.1016/j.icarus.2005.04.001
R.W. Potter, J.W. Head, D. Guo, J. Liu, L. Xiao, The Apollo peak-ring impact basin: insights into the structure and evolution of the South Pole-Aitken basin. Icarus **306**, 139–149 (2018). https://doi.org/10.1016/j.icarus.2018.02.007
S.R. Schwartz, P. Michel, M. Jutzi, S. Marchi, Y. Zhang, D.C. Richardson, Catastrophic disruptions as the origin of bilobate comets. Nat. Astron. **2**, 379–382 (2018). https://doi.org/10.1038/s41550-018-0395-2
K. Sugiura, S. Inutsuka, An extension of Godunov SPH: application to negative pressure media. J. Comput. Phys. **308**, 171–197 (2016). https://doi.org/10.1016/j.jcp.2015.12.030
K. Sugiura, S. Inutsuka, An extension of Godunov SPH II: application to elastic dynamics. J. Comput. Phys. **333**, 78–103 (2017). https://doi.org/10.1016/j.jcp.2016.12.026
K.J. Walsh, D.C. Richardson, Binary near-earth asteroid formation: rubble pile model of tidal disruptions. Icarus **180**(1), 201–216 (2006). https://doi.org/10.1016/j.icarus.2005.08.015
K.J. Walsh, D.C. Richardson, P. Michel, Rotational breakup as the origin of small binary asteroids. Nature **454**, 188–191 (2008). https://doi.org/10.1038/nature07078

The manufacturer's authorised representative in the EU is Springer Nature Customer Service Centre GmbH, Europaplatz 3, 69115 Heidelberg, Germany. If you have any concerns regarding our products, please contact ProductSafety@springernature.com

Printed and bound by CPI Group (UK) Ltd, Croydon, CR0 4YY
30/09/2026
02225574-0007